Synthesis Lectures on Engineering, Science, and Technology

The focus of this series is general topics, and applications about, and for, engineers and scientists on a wide array of applications, methods and advances. Most titles cover subjects such as professional development, education, and study skills, as well as basic introductory undergraduate material and other topics appropriate for a broader and less technical audience.

Rajan Chattamvelli · Ramalingam Shanmugam

Random Variables
for Scientists and Engineers

 Springer

Rajan Chattamvelli
School of Computer Science and Engineering
Amrita University
Amaravati, Andhra Pradesh, India

Ramalingam Shanmugam
School of Health Administration
Texas State University
San Marcos, TX, USA

ISSN 2690-0300   ISSN 2690-0327 (electronic)
Synthesis Lectures on Engineering, Science, and Technology
ISBN 978-3-031-58930-0   ISBN 978-3-031-58931-7 (eBook)
https://doi.org/10.1007/978-3-031-58931-7

This Springer imprint is published by the registered company Springer Nature Switzerland AG
The registered company address is: Gewerbestrasse 11, 6330 Cham, Switzerland

Paper in this product is recyclable.

Preface

This book gives an introduction to random variables (RVs) and their transformations. The aim is to give a clear exposition of mathematical expectations, univariate RVs, and joint distributions.

Chapter 1 introduces RVs and mathematical expectations. Discrete and continuous random variables and their basic properties are discussed. Cumulative distribution functions (CDF), survival functions (SF), quantile functions (QF), and their properties are detailed. Arithmetic operations on RVs give rise to new RVs. A simple technique to find the expectation of functions of RVs is given. This is followed by a discussion on moments (ordinary, central, factorial), variance, and co-variance as expected values. The conditional expectation is introduced and its applications to finding the moments of infinite mixture distributions such as noncentral chi-square, noncentral beta, and noncentral F distributions are demonstrated. Chapter 1 ends with a discussion on the applications of random variables.

Chapter 2 discusses the distributions of functions of single RVs. Topics discussed include distribution of absolute value, method of distribution functions (MoDF), change of variable technique (CoVT), distribution of sums, squares, square-roots, reciprocals, trigonometric, and transcendental functions, minimum and maximum, integer and fractional parts, arbitrary functions, and ratio of sums. A summary table of common single-variable transformations is provided in Sect. 2.11.1. These results are used to express the mean deviation of continuous distributions as a simple integral from lower limit to F(mean) where F() is the CDF. The chapter ends with a discussion of transformations of normal variates and some applications of functions of random variables in various fields.

Distribution of functions of several random variables is introduced in Chap. 3. Marginal and conditional distributions are briefly discussed. The Jacobian of matrix transformation is described and its applications in various fields are cited. This is illustrated in finding the distribution of a variety of transformations including products, ratios, and nonlinear functions of two or more RVs. A "do-little" technique to quickly find the Jacobian of transformation of random variables useful in statistics is described. Plane-polar, spherical-polar, cylindrical-polar, toroidal-polar, Helmert, and Rosenblatt transformations are also discussed. Integral calculus is heavily used in this chapter. A summary table of common transformation in two variables is provided in Sect. 3.4. The chapter ends with a discussion on copula-based methods.

Suggestions for changes are always welcome. For any suggestions on improvement please contact rajancv@am.amrita.edu.

Amaravati, India   Rajan Chattamvelli
February 2024   Ramalingam Shanmugam

Contents

About the Authors

Rajan Chattamvelli is a professor in the school of computer science and engineering at Amrita University, Amaravati. He has published more than 20 research articles in international journals of repute and at various conferences. His research interests are in computational statistics, design of algorithms, parallel computing, cryptography, data mining, machine learning, and big data analytics. His prior assignments include Denver Public Health, Colorado; Metromail Corporation, Lincoln, Nebraska; Frederick University, Cyprus; Indian Institute of Management; Periyar Maniammai University, Thanjavur; Presidency University, Bengaluru; and VIT University, Vellore.

Ramalingam Shanmugam is an honorary professor in the school of Health Administration at Texas State University, San Marcos. He is the editor of the journals: *Advances in Life Sciences, Global Journal of Research and Review, International Journal of Research in Medical Sciences, Kenkyu Journal of Epidemiology and Community Medicine*, and *Journal of Obesity and Metabolism* and book-review editor of the *Journal of Statistical Computation and Simulation*. He has published more than 200 research articles and 120 conference papers. His areas of research include theoretical and computational statistics, number theory, operations research, biostatistics, decision making, infectious disease modeling, patient risk management, cost-effective analysis, and epidemiology. His prior assignments include University of South Alabama, University of Colorado at Denver, Argonne National Labs, Indian Statistical Institute, and Mississippi State University. He is the president of the San Antonio chapter of the *American Statistical Association* and a fellow of the International Statistical Institute.

List of Tables

Mathematical Expectation

1

This chapter introduces random variables and mathematical expectation. Discrete and continuous random variables and their basic properties are discussed. A simple technique to find the expectation of functions of random variables is given. This is followed by a discussion on moments (ordinary, central, factorial) and variance as expected values. The conditional expectation is introduced and its applications to finding the low-order moments of mixture distributions are demonstrated. A brief discussion of inverse and incomplete moments, and distance as expected value follows it. The chapter ends with some common applications of random variables.

1.1   Meaning of Expectation

The concept of "expected value" appeared for the first time in the works of Christian Huygens (1629–1695) around 1657. It was used to predict the possible gains in gambling and games of chance. It can be associated either with a single random variable (RV) or a well-defined function of the RV. Location measures (such as the mean, median, mode) condense the information in a sample as a single number (in univariate case). Analogous measures are needed to succinctly present the characteristics of statistical populations or random experiments. This is where the concept of expectation comes in. The functional form of the population is known precisely in most of the discussions below. But theoretically, the concept is valid even when the exact form is either unknown or is partially known (as in random experiments involving circuits, transmission medium, moving objects, etc.) The expected value of observed phenomena are applicable in the long-run during which an event of interest is going to occur repeatedly under identical experimental conditions. This may sometimes be observed from past data. For instance, consider the price of a stock that fluctuates randomly

© The Author(s), under exclusive license to Springer Nature Switzerland AG 2024   1
R. Chattamvelli and R. Shanmugam, *Random Variables for Scientists and Engineers*,
Synthesis Lectures on Engineering, Science, and Technology,
https://doi.org/10.1007/978-3-031-58931-7_1

over time. We could find the expected value of stock price by averaging the observed values over a suitable time period using an uncertainly measure (probability for it to rise, fall or remain steady). Insurance companies use expected values to predict various quantities. For example, flooding and power outages are quite common during monsoon season in tropical cities with poor drainage facilities. If a weatherman predicts that a heavy thunderstorm is likely in the next few days with a probability of 0.90, the insurance companies can use this information using past data to estimate the expected amount on insurance claims that will be received after the event. This is discussed further in Sect. 1.3. The notion of mathematical expectation (or simply called expectation) relies on one or more *random variables* defined below.

1.2   Random Variables

The concept of RV is of prime importance in mathematical expectation. It is defined on the sample space of a random experiment, which is an experiment that can be repeated any number of times under (more or less) identical conditions.

Definition 1.1   The set of all possible outcomes of a random experiment is called the sample space. It is usually denoted by the Greek letter Ω or the letter S.

The outcome of a random experiment can be given names, labels or an enumeration. Thus when a coin is tossed, the possible outcomes are represented as {H, T}, {Head, Tail} or simply as {0, 1} where 0 denotes the nonoccurence and 1 denotes the occurrence of an event.

Definition 1.2   A random variable is a function defined on the sample space of a random experiment that maps each possible outcome of the sample space to real numbers such that the associated probabilities sum to one.

Mathematically, an RV is a rule that assigns a unique numerical value to each event (outcome) of a random experiment (Fig. 1.1). An indicator function is a special type of RV in which each element in the sample space is mapped to either 0 or 1. If E is an arbitrary event

Fig. 1.1 Random variable
maps sample space to real line

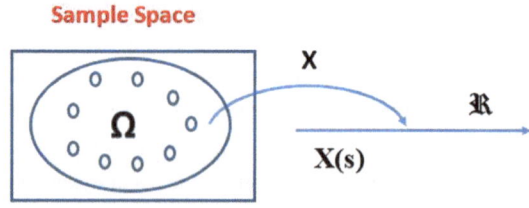

$$I_E(s) = \begin{cases} 1 \text{ if } s \in E \\ 0 \text{ if } s \notin E. \end{cases}$$

1.2.1   Realisation of Random Variables

An RV ensues when (i) an experimenter performs a random experiment, (ii) defines a random experiment on a hypothetical experiment (like tossing a die or coin), (iii) observes the results of an experimental outcome, or (iv) a physical or natural process generates data that can be approximated by a statistical law. Such occurrences are denoted by lowercase letters (in which case it is an implicit assumption that the corresponding uppercase letter denotes the RV). RandVar is an R-package to implement random variables. (r-project.org)

Random processes are RVs where the values of a variable vary systematically over time. Note that the outcomes can be anything including numbers, labels, symbols or even text strings. Thus in an industrial experiment that checks whether a machine or part is defective, the outcomes could be {DD, DN, ND, NN} where 'D' denotes defective or non-working and 'N' denotes non-defective or in good working condition. There are multiple ways in which these outcomes can be mapped numerically. If the aim of a study is to identify defectives, we could map 'D' to a '1' and 'N' to a '0' so that the possible values the random variable can take are {2, 1, 0}. However, some people prefer to always map a defective to a zero and non-defective to a one. In this case the probabilities simply get reversed. An RV can be denoted as $X{:}\Omega \rightarrow \mathbb{R}$ where the values that it takes in $\mathbb{R}$ are known from the mapping used (see Fig. 1.1). All RVs are denoted by uppercase letters and particular values by lowercase letters in the following discussion.

We will denote "distributed as" by the symbol "∼" (which is the *tilde* symbol), and 'independently and identically distributed' as IID. Abbreviations will be used for distributions in an unambiguous way (POIS for Poisson, CUNI for continuous uniform, DUNI for discrete uniform, EXP for exponential, BINO(n, p) for binomial, etc.). Thus $X{\sim}\text{CUNI}(0, 1) = U(0, 1)$ is read as "X is distributed as continuous uniform in [0, 1]".

1.2.2   Discrete and Continuous Random Variables

An RV can be discrete, continuous or mixed type. Among these discrete RVs are often used with count data, and continuous RVs are used when measurements are done by machines or computed using mathematical formula (like BMI).

Definition 1.3 An RV is discrete if the set of possible values (outcomes) that it could take is finite or countably infinite. Mathematically, X is discrete if $x \in \{x_1, x_2, \ldots, x_n\}$ or $x \in \{x_1, x_2, \ldots, \}$. This concept is easy to understand for discrete RVs as the number of events in the sample space (domain) is countably finite. The domain can also be a part of an integer (such as half-integer). As an example, suppose X represents the number of medical leaves taken by an employee where the employer allows either full-day leave or half-day leave types only. In this case the domain can be integers or half-integers per month. The domain of several discrete RVs are an ordered sequence. Consider a vehicle insurance company who screens new customers for the number of past accidents. A great majority of customers might have no severe accidents at all (x = 0). There could be several customers with one accident (x = 1), two accidents (x = 2) and so on. If the number of applicants is large, we expect x to take values in (0, 1, . . ., m) where m is a fixed number. However, there could be breaks in such a sequence in which case we simply insert missing values and assume the corresponding frequency to be zero. This makes it into an ordered sequence.

In some practical applications, the discrete RV represents *count* data that may occur over time or space. As examples, the number of defective items in a shipment, number of patients with a particular symptom, number of alpha-particles emitted by a radioactive source in a small time-interval, number of crossovers that have occurred in a genome, number of earthquakes in a particular locality are all examples of count data. Here the last two examples are occurrences over space while others occur over time. Any number of RVs can be defined on a given sample space. These may be related or independent. There are some numerical values associated with each outcome of a random experiment. For instance, consider the toss of a die with six faces. The possible outcomes are {1, 2, 3, 4, 5, 6}, each with probability 1/6. If X denotes the face that turns up, we express it mathematically as $f(x) = 1/6$ for x = 1, 2, . . . 6. We have simply assigned a mathematical function to each outcome of a random experiment. This experiment is called equally likely (and the RV as equiprobable) because the associated probabilities are all equal. The corresponding mapping from the sample space to the real line is called probability mass function (PMF) in the discrete case and probability density function (PDF) in the continuous case. This is the most common way to define a discrete RV. Similarly, $f(x) = q^x p$ is a mathematically defined RV associated with a countably infinite sample space consisting of the sequence 0, 1, 2, . . ., ∞ associated with a random experiment (with two mutually exclusive outcomes) that is repeated until a special event occurs. The above RV occurs when a coin is tossed repeatedly until a Head (with $\Pr(Head)$ = p) occurs. The values of a well-defined RV are subject to chance (i.e. it is stochastic). Thus even if we toss a coin thousands of times, we can't predict in advance what the next outcome is going to be. Some RVs are time-dependent. Consider the number of phone calls passing through an automated telephone exchange in a fixed time interval (say 10 s). This varies from time to time, but could be modeled using a statistical law if the time interval is properly chosen. Similarly, the number of phone calls received at an emergency number or fire-station can be modeled using a statistical law such as the Poisson law. As these are rare events, the time interval is large.

There is one more way to define discrete RVs. It is called complete enumeration method. Consider the RV $p(1) = 0.2, p(2) = 0.6, p(3) = 0.2$. Here x takes 3 values {1, 2, 3}. It is a well-defined RV as the probabilities add up to one. This can also be written as $p(x = 1) = 0.2, p(x = 2) = 0.6, p(x = 3) = 0.2$ for a univariate RV X. This notation can be extended to bivariate and higher-dimensional RVs. It is better suited when the sample space is of small size. A special case is the degenerate RV in which there is just a single value x = c with associated probability 1 (also called point mass or degenerate distribution). The individual probabilities can also be defined using a recurrence relation (Chap. 2, Sect. 2.12.2). When the sample space is infinite, either the mathematical notation or recurrence relation is better suited to describe the RV.

Two or more discrete RVs can be compared using equality or relational operators ($<$, $\leq$, $>$, $\geq$). Two RVs X and Y are equal (i.e. X = Y) iff X(s) = Y(s) $\forall s \in \Omega$ (i.e. $p_x(k) = p_y(k)\forall k$). Similarly X$\leq$Y if X(s)$\leq$Y(s), and X$\geq$Y if X(s)$\geq$Y(s) $\forall s \in \Omega$. If X and Y are defined on the same sample space, X–Y is a well-defined RV so that X$\leq$Y$\equiv$ (X–Y) $\leq$ 0. In particular, X$\leq$c means that X(s) $\leq$ c $\forall$ s$\in \Omega$. We can also use lower bounds (X $\geq$ c) or both bounds (c $\leq$ X $\leq$ d) on RVs.

The second type of RV is the continuous RV, in which the possible values are any number on the real line (not necessarily integer or evenly spaced values) or a closed curve (such as the points on the circumference of a circle).

Definition 1.4 An RV is continuous if it can take values on a continuous scale x: $a < x < b$ where a, b are any real numbers (a, b $\in$ **R**), (including ∞).

Convolution of continuous RVs results when the intervals are disjoint. In most practical applications, the continuous RV represents measured data either in a unit or as a percentage. The measurement can be done either using devices or machines (e.g.: body temperature or blood pressure; speedometer reading of a vehicle), computed using mathematical formula (e.g.: body mass index (BMI) defined as BMI = weight in Kg./(height in meter squared), Cardiac Index (CI) defined as CI = CO/BSA, CO = cardiac output (measured in cubic ml) and BSA = total body surface area (measured in cm^2)).[1] The unit of the variable is needed when data are graphed but can often be changed from one unit to another. Thus the body temperature can be expressed in Celsius or Fahrenheit. The body temperature is an interval-scale variable because zero degrees does not tally in different scales (0 °C = −32 °F), but differences between temperatures make sense [Chattamvelli (2016)]. Several clinical variables are ratio-measures in NOIR typology (WBC count, Hemoglobin level (Hb), Uric acid level).[2] Two or more continuous RVs can be compared using equality or relational operators ($<$, $\leq$, $>$, $\geq$). If X and Y are continuous and defined on the same sample space, X–Y is a well-defined RV so that X$<$Y$\equiv$ (X–Y) $\leq$0. This is easy to find when X and Y are independent. In particular, X$\leq$c means that Y = (X–c) $\leq$0. We can also use lower bounds (X$\geq$c) or both bounds (c$\leq$X$\leq$d) on RVs. Bounds can also be placed on functions of RVs. For example Pr[$X^2 \leq$k] = Pr[$-\sqrt{k} \leq X \leq \sqrt{k}$].

---

[1] As the numerator unit is ml^3 and denominator unit is a square, the units do not cancel out, so that CI is an absolute measure. This could be converted into a relative measure by dividing it by height.

[2] Borg's category ratio-scale (CR-10) is a common scale used in medical sciences in which 0 indicates total absence, 0.5 = very very weak, 1 = very weak 2 = weak (light), 3 = moderate, 4 = somewhat strong, 5 = strong, 7 = very strong, 10 = very very strong (max).

Definition 1.5 A probability mass function defined on the discrete sample space of a random experiment is a mapping that can be represented as an ordered pair {x, f(x)} if for each possible outcome x of the sample space, the following three conditions are satisfied:— (i) f(x) ≥ 0 ∀ x values, (ii) $\sum_x$ f(x) = 1, (iii) Pr($X = x$) = f(x) unambiguously.

It is sometimes called a probability function or frequency function. Note that in the case of univariate continuous RVs, it is the area in an infinitesimal interval $\int_{x-\epsilon}^{x+\epsilon} f(x)dx$ where $\epsilon = dx/2$ that represents the probability value of the variable at x. This becomes infinitesimal volumes in higher dimensions. Hence Pr(x = c) could be any non-negative number for a fixed c in the continuous case. One exception is the continuous uniform RV defined on [0, 1] as f(x) = 1 for any value of x in its range 0≤ x ≤1.

Definition 1.6 A probability density function defined on the continuous sample space of a random experiment is a mapping that can be represented as {x, $\int_{-dx/2}^{dx/2}$f(x)dx} satisfying the following conditions:— (i) f(x) ≥ 0 ∀ x values, (ii) $\int_{x=-\infty}^{\infty} f(x)dx = 1$, (iii) Pr($a < X < b$) = $\int_a^b f(x)dx$.

1.2.3 Numeric Value of PDF

To check if a given PMF is well-defined, we have to check if $p(x_i) \geq 0$, $p(x_i) \leq 1$, and $\sum_i p(x_i) = 1$ in the discrete case. The summation is replaced by integration in the continuous case. The condition $p(x_i) \leq 1$ need not always hold in the continuous case, however. In other words, the numeric value of PDF evaluated at some points may exceed one in magnitude for some continuous distributions. This means that the PDF value at a particular point can't be interpreted as probability in continuous case (see Fig. 1.2). One example is the PDF of the sample Pearson's correlation coefficient for $\rho \neq 0$ which can far-exceed one because it is highly skewed when $\rho \to \pm 1$, and degenerates in the limiting case, where PDF can approach infinity! [Chattamvelli (2024)]. Another example is the square-root of U(0, 1)

Fig. 1.2 PDF value can't be interpreted as probability

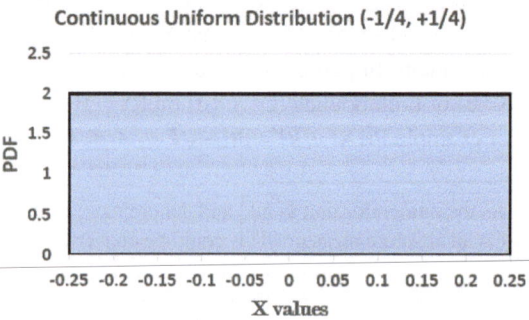

variate ($Y = \sqrt{X}$ where $X\sim U(0, 1)$), whose PDF can exceed one when y > 0.50. In general, if a continuous uniform distribution (CUNI) is defined on (0, k) where k<1, or on (−k, +k) where k<1/2, the PDF will exceed one in magnitude. Thus, if the diameter of a manufactured product can deviate from the ideal value in both directions uniformly in (−0.25, 0.25) units, the PDF is f(u) = 1/[(0.25−(−0.25)] = 1/0.5 = 2 for all x values in (−1/4, +1/4) (Fig. 1.2). Nevertheless, the integrated value in a small interval will be in [0, 1), although theoretically it is defined as f(x) = $\int_{t=(x-1/n)}^{x} f_X(t)dt$ as n becomes large so that in the limiting case it becomes zero. This could happen not only for univariate RVs, but in higher dimensions as well.

Check if p(x) is a PMF

Check whether the RV defined as $\Pr(x = 0) = q^2$, $\Pr(x = 1) = 2pq$, $\Pr(x = 2) = p^2$ is a PMF where $0 < p < 1$ and $p + q = 1$. Find its mean and variance.

Solution As $0 < p, q < 1$, all probabilities are positive. Also $q^2 + 2pq + p^2 = (q + p)^2 = 1$ showing that the probabilities add up to 1. Hence it is a PMF (Fig. 1.3). The mean is given by E(X) = $0 \times q^2 + 1 \times 2pq + 2 \times p^2 = 2p(p + q) = 2p$. Similarly, $E(X^2) = 0^2 \times q^2 + 1^2 \times 2pq + 2^2 \times p^2 = 2pq + 4p^2$. Hence V(x) = $E(X^2) - [E(X)]^2 = 2pq + 4p^2 - (2p)^2 = 2pq$.

1.2.4   Arithmetic on RVs

If two or more RVs are defined on the same sample space (compatible RVs), we could form new RVs using some arithmetic operations like addition, subtraction, multiplication and powers. Thus if X and Y are compatible RVs, U = X + Y, V = X−Y, W = XY, Z = X² could all be RVs. This may not always result in a meaningful or useful RV, however. As an example, the square of a standard normal RV is chi-square distributed, whereas the square

Fig. 1.3 PMF checking example

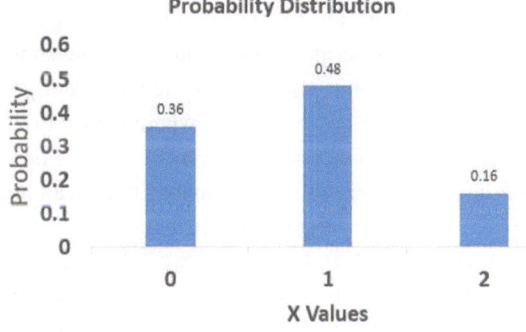

of a Cauchy distribution is seldom used in practice. Similarly, when X and Y are positive RVs, the difference U = X–Y takes negative values and may not always be useful.

The shape of the distribution will change when a change of scale transformation is applied, but shape will remain the same and location only will differ when a change of origin transformation is applied.

Consider the BMI of people in a group. This is always positive and with a clear lower limit (that could vary from one group to another). The BMI of adults are almost always greater than 15. As it is a ratio, it could take any decimal value (although for medical studies it is truncated to nearest integer). Hence it is a continuous variable. There are many other ratio measures (also called metric scales which are the most informative scale) used in medical sciences; load-ratio, diffusion-ratio, etc. used in engineering fields. Examples are the waist-by-hip ratio, and waist-to-height ratio.

Find the unknowns c and d

If $X \sim$ CUNI(a, b), find the unknown constants c, d such that $Y = cX + d$ is CUNI(0, 1) (i.e. U(0, 1)) distributed.

Solution The PDF of X is f(x; a, b) = 1/(b–a), a<x<b. The mean and variance of X are respectively (a+b)/2 and $(b-a)^2/12$. Similarly, the mean and variance of U(0, 1) are 1/2 and 1/12 respectively. First consider the variance of Y as Var(Y) = c^2 Var(X). Substitute the values to get $1/12 = c^2(b-a)^2/12$, which gives c = ±1/(b–a). As E(y) = c E(X) + d, we get 1/2 = ±1/(b–a) (a+b)/2 + d, so that d = –a/(b–a) or b/(b–a).

1.2.5   Cumulative Distribution Function

The cumulative probabilities (in discrete case) are computed by summing individual probabilities from the lowest possible x value to a higher number. The implicit assumption here is that the RV is arranged in increasing order of possible values of the outcomes. Symbolically $F_x(x) = \Pr[X \leq x]$. It is usually denoted by capital English letters like F(), G(), etc. with or without a subscript (that indicates variable names or labels; or degrees of freedom (DoF) in the case of some distributions such as T and χ^2). Thus the CDF in discrete case can be written as $F_x(k) = \sum_{X \leq k} p(x)$ where p(x) = $\Pr(X = x)$. It is a function from Ω to [0, 1] (F:$\Omega \rightarrow$ [0, 1]). For x = 1, F(x) = p(x) = p(1). For x = 2, F(x) = p(1) + p(2), and so on. Sum of the probabilities in any continuous range can be written in terms of CDF as $\Pr[a<X \leq b]$ = F(b)–F(a). The cumulative distribution function (CDF) is a jump- (or step-) function if X is discrete (Fig. 1.4). In this case we call it a cumulative probability function. Analogously the sum of right tail probabilities is called the survival function (or survival probabilities) S(x). They are related as F(x) = 1–S(x). It satisfies the following properties for a discrete RV.

Fig. 1.4 CDF of discrete RV

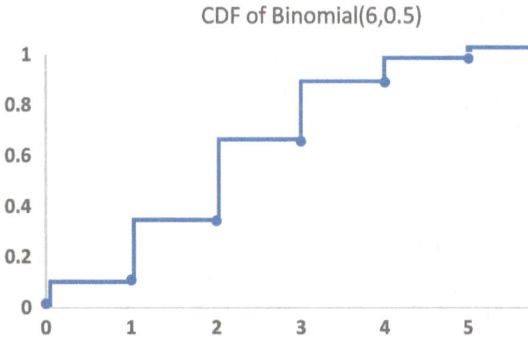

CDF of Binomial(6,0.5)

1. $\Pr(F(x) \le u) = u$, $\Pr(F(x) \ge u) = 1-u$ (continuous case).
2. $\Pr(a < x < b) = \Pr(x \le b) - \Pr(x \le a) = F(b)-F(a)$.
3. $\Pr(a \le x \le b) = \Pr(x \le b) - \Pr(x < a) = F(b)-F(a)+p(a)$.
4. $\Pr(a < x < b) = \Pr(a < x \le b) - \Pr(x = b) = F(b)-F(a)-p(b)$.
5. $F(F^{-1}(x)) \ge x$, and $F(x) \ge u$ iff $x \ge F^{-1}(u)$ where equality holds in the continuous case.
6. If $F_1(x)$, $F_2(x)$, ... $F_k(x)$ are CDFs and $p_1, p_2, \ldots, p_k$ are real numbers such that $\sum_{j=1}^{k} p_j = 1$, then $\sum_{j=1}^{k} p_j F_j(x)$ is also a CDF (called a mixture distribution).

Integration instead of summation is used when X is continuous (see Table 1.1). Thus irrespective of the nature of the RV, we can define the CDF as the probability that the RV (say X) takes values less than or equal to x, where x is a specified number within the range. Obviously the CDF is an increasing function of x. This means that $F(x)-F(x-1) = p(x)$ for discrete RVs. In the case of continuous RVs, we have (i) $F(x)$ is a non-decreasing function of x, (ii) $\lim_{x \to ll} F(x) = 0$ (i.e. F(ll) = 0), (iii) $\lim_{x \to ul} F(x) = 1$ (i.e. F(ul) = 1), and (iv) $\frac{\partial}{\partial x} F(x) = F'(x) = f(x)$ where ll and ul are the lower and upper limits, (v) $F(x)$ is continuous function of x on the right, with countable number of discontinuities, if any. Property (i) follows trivially due to our implicit assumption that the outcomes of the random experiment are arranged in ascending order of their values. This means that if x_1 is strictly less than x_2, then all sample points that are part of the left tail up to x_1 are automatically part of the left tail up to x_2. Thus the sum of probabilities up to x_1 is strictly less than that up to x_2. Other properties follow easily due to the increasing nature of $F(x)$, which must eventually equal 1.

As the CDF accumulates probabilities from the left tail, it easily follows that for b>a,

$$F(b) = \begin{cases} F(a) + \sum_{k=a+1}^{b} p(k) & \text{if } X \text{ is discrete;} \\ F(a) + \int_a^b f(x)dx & \text{if } X \text{ is continuous.} \end{cases}$$

For symmetric RVs that extends to the negative side, $F(\mu+x) = 1-F(\mu-x)$. The PDF of a continuous distribution can be obtained from the CDF by differentiation, and that of a discrete distribution can be obtained from the CDF by differencing. Symbolically,

Table 1.1 Comparison of discrete and continuous RVs

Property	Discrete	Continuous	Comment
PMF f(x)	Prob. mass function	Prob. density function	
CDF $F_X(x)$	$\sum_{k=lo}^{x} f(k)$	$\int_{u=lo}^{x} f(u)\, du$	lo is the lower limit
Sum	$\sum_{x=lo}^{hi} f(x) = 1$	$\int_{x=lo}^{hi} f(x)\, dx = 1$	lo = lower, hi = upper limit
Partial sum	$\sum_{k=a}^{b} f(x) =$ F(b)–F(a)	$\int_{a}^{b} f(x) dx = $ F(b)–F(a)	
f(x) from F(x)	f(x) = F(x)–F(x–1)	$f(x) = \partial F(x)/\partial x$	f(lo) = F(lo) discrete
$x_1 < x_2$	$F(x_1) \leq F(x_2)$	$F(x_1) < F(x_2)$	

For the continuous case, f(x = c) could be anything ≥ 0 for fixed c. It is the area in an infinitesimal interval $\int_{x-\epsilon}^{x+\epsilon} f(x) dx$ where $\epsilon = dx/2$ that represents the probability at x (see Fig. 1.5)

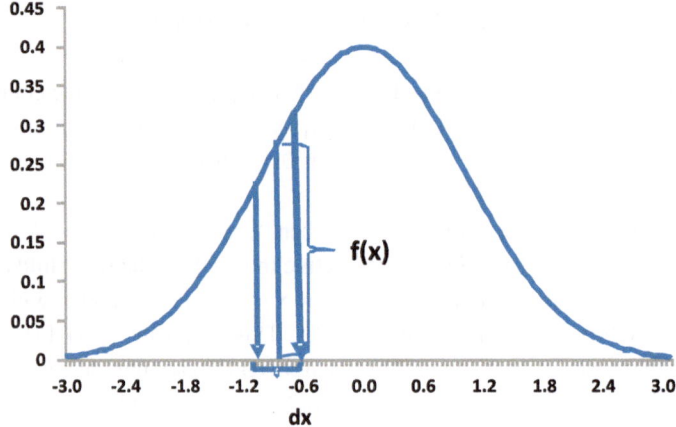

Fig. 1.5 Continuous PDF

$$f(x) = \lim_{\Delta x \to 0} [F(x + \Delta x) - F(x)]/\Delta x = \partial F(x)/\partial x, \qquad (1.1)$$

provided the limit exists.

Problem 1.1 Check if F(x) = $\frac{2}{\pi} \sin^{-1}(\sqrt{x})$ for 0< x <1 is a CDF.

Problem 1.2 Check if F(x) = $\frac{1}{2} + \frac{1}{\pi} \tan^{-1}(x)$ for $-\infty < x < \infty$ is a CDF.

$\Pr[c-\delta < x < c+\delta]$ as $F(x)$

Express $\Pr[c-\delta < X < c+\delta]$ in terms of CDF. For what value of c is this area maximum for bell-shaped curves?.

Solution Write $\Pr[c-\delta < X < c+\delta]$ as $F(c+\delta)-F(c-\delta)$. This is maximum for $c = \mu$ for bell-shaped distributions because the given expression can be written as $P|x - c| < \delta$. As δ is a constant, this is maximum for $c = \mu$.

For zero-symmetric distributions, $F(-x) = 1-F(x)$ so that $F(x) + F(-x) = 1$. Similarly, $F(k) - F(-k) = \int_{-k}^{k} f(x)dx = 2 \int_{0}^{k} f(x)dx$. In general, if a distribution is symmetric around c, then $F(c-x) + F(c + x) = 1$.

Check if $F(x)$ is a CDF

Verify whether $F(x; c, d) = (1/(d-c))[dx^c-cx^d]$ over [0, 1] is a CDF.

Solution Differentiate wrt x to get $f(x; c, d) = (1/(d-c))[cdx^{c-1}-cdx^{d-1}]$. Take cd as a common factor to get $f(x; c, d) = (cd/(d-c))[x^{c-1} - x^{d-1}]$. As $c<d$, and $0 \le x \le 1$, the expression inside the square bracket is always positive. This shows that $F(x)$ is a non-decreasing function of x. As $ll = 0$, $F(ll) = F(0) = (1/(d-c))[0-0] = 0$, showing that $\lim_{x \to ll} F(x) = 0$. As $ul = 1$, $F(ul) = F(1) = (1/(d-c))[d-c] = 1$, showing that $\lim_{x \to ul} F(x) = 1$. Obviously, $\int_{0}^{1} f(x; c, d)dx = (cd/(d-c)) [x^c/c-cx^d/d]|_{0}^{1} = (cd/(d-c))[1/c-1/d] = 1$. Here $F(x)$ does not have discontinuities in the interval [0, 1]. Hence all the conditions mentioned above are satisfied, proving that $F(x)$ is indeed a CDF.

If $F(x)$ denotes the CDF of a distribution defined on the interval $(0, 1)$, then $F^{-1}(x)$ is also a CDF of a well-defined RV (called complementary distribution) defined on the same interval. For example, the beta-I distribution is a RV defined on the unit interval with CDF as the incomplete beta function, which is invertible. Hence $F^{-1}(x)$ is the CDF of complementary beta distribution [Jones (2002)].

1.2.6   Quantile Function

Quantile function is the inverse of CDF in which we seek a value of $x = k$ (say) such that $\Pr[X \le k] = p$ where p is a probability in [0, 1] (i.e. x: $F^{-1}(x) = p$). It is denoted by $Q(p)$, $0 \le p \le 1$, and can take any value within the range of X. It can be defined in terms of CDF as $Q(p) = \min_{x \in \Omega} F(x) \ge p$. In other words, it is the smallest value of x that satisfies $F(x) \ge p$. For continuous RVs $Q(p) = F^{-1}(x)$: $F(x) = p$. This is easy to find for univariate

RVs, but may not be easy for higher dimensional RVs. When F(x) is invertible, one could find the value of $f(F^{-1}(x))$, which is known as the density quantile function. Quartiles are the most common quantiles in which p is a multiple of 1/4. The first quartile Q_1 is obtained for p = 1/4, and the third quartile Q_3 for p = 3/4. For symmetric distributions, $(Q_1 + Q_3)/2 = Q_2 = Q_1 + (Q_3 - Q_1)/2$ is the second quartile (which is the median). Deciles divide the entire interval into multiples of 1/10 (of the area or total probability) and are denoted as D_k for k = 1, 2, ...9. Similarly, 99 percentiles divide the total probability into 100 equal parts.

1.3   Mathematical Expectation

The expected value can be thought of as the arithmetic mean (weighted average) of RVs or populations, where the weights are the probabilities.

Definition 1.7 Mathematical expectation is used to concisely quantify the mean value of an event, an experiment, an RV or a function of it in a population of interest.

It is also called the expected or average value of the argument, or the center of mass in physical sciences. It can be any real number within the range for real-valued RVs. The expected value of integer-valued RVs need not be an integer. It is so called because (i) population may be unknown, (ii) population could be un-enumerable, (iii) it could be a function of the unknown parameters of the population, and (iv) it is just the *expected value* (and not the exact value) of the population under study. It is defined for discrete as well as continuous RVs, and any well-defined functions of them. It is a scalar for univariate populations, and a vector for multivariate RVs. It is denoted as E(X), E[X] or E X, where E is the *expectation operator*, followed by the argument. The argument (also called the operand, which is usually an expression or a function in capital letters) of "E" can be any well-defined function of X, including integer and fractional powers of X, or conditional distributions. Positive integer powers of X are called ordinary moments, negative integer powers are inverse (or negative) moments, and fractional powers are fractional moments. All moments of a distribution need not exist, however. One example is the Cauchy distribution.

The argument of expectation operator can also be another expectation expression. Thus the "E" operator can be recursively nested (Sect. 1.5) (this is called the law of iterated expectations). In the rest of the chapter, we will simply call it the expected value. In the particular case when the argument is X itself, it is called the population mean (first moment).

Definition 1.8 The expected value of a univariate RV is defined as:

$$E(x) = \begin{cases} \sum_{k=-\infty}^{\infty} x_k p_k & \text{if } X \text{ is discrete;} \\ \int_{-\infty}^{\infty} xf(x)dx = \int_{-\infty}^{\infty} x \, dF(x) & \text{if } X \text{ is continuous.} \end{cases}$$

Whenever this sum is *absolutely convergent*, we call it the expected value of X. This means that $\int |x| dF(x) < \infty$ in the continuous case, and $\sum_x |x| f(x) < \infty$ in the discrete case. The continuous case can be considered as the limiting form of the discrete case as $E(X) = \lim_{N \to \infty} \frac{1}{N} \sum_i X_i$, which means that if a random experiment is repeated many times under identical conditions and the value x_i of the outcome is measured each time, the expected value of X is the mean of the limiting value as N becomes large. As the binomial distribution takes values between 0 and a positive integer n, all expected values of binomial RVs use the summation range from 0 to n. Similarly, as the Poisson and geometric distributions take values between 0 and ∞, the summation is carried out in this range only. Hence the very first step in finding the expectation is to figure out the exact range.

As p'_ks are probabilities that sum to 1 ($\sum_k p_k = 1$, $\int f(x) dx = 1$), the above sum will almost always converge when $\sum_{k=-\infty}^{\infty} |x_k| p_k < \infty$ or $\int_{-\infty}^{\infty} |x| f(x) dx < \infty$. On occasion this sum may diverge, in which case we say that the expected value does not exist. This seldom happens for RVs defined over a finite range.

As mentioned above, E(X) is a weighted average of all possible values that the variable can take with corresponding probabilities as the weights. In other words, the notation E(X) can be considered as an *expectation operator*, operating upon the entire range of its argument (which is not shown, but is understood from context). The argument of this expectation operator is quite often a RV, with the implied meaning that it is the centroid of the argument. As discussed above, the argument could also be any well-defined function of the RV (e.g.: $E(X^2)$, $E(X^{-k})$). If all weights (p_k or $f(x_k)$) are equal (and the range of X is finite), the expected value reduces to the ordinary average (arithmetic mean) of all possible values. For instance, let X take values in the range $1 \ldots$ n and let $p_k = 1/n$ (so that $\sum_k p_k = 1$) then E(X) $= \sum_k p_k x_k = \sum_k k/n = \frac{1}{n} \sum_k k = n(n+1)/[2n] = (n+1)/2$. When all probabilities are not equal, $E(X) = \lim_{n \to \infty} \bar{x}_n$, which means that the sample mean converges to the population mean in the limiting case when n becomes very large.

Expected value of c/x^2, x ≥ 1

Find the expected value of the RV X with PDF f(x) = c/x^2, x ≥ 1 where c is the normalizing constant.

Solution As the total probability must be one, $c \int_1^\infty 1/x^2 \, dx = 1 \Rightarrow -c[1/x|_1^\infty] = 1$ giving c = 1. The expected value is E(X) = $\int_1^\infty (1/x) dx = [\log(x)|_1^\infty] = \infty$. Hence E(X) does not exist.

In many of the examples given below, we are interested in the expected value of random experiments. In such situations, E(X) can be considered as the average value of the experiment, if it is repeated under identical conditions a large number of times.

Find normalising constant

Find K to make the following functions a PDF. Then find E(X) and $E(X^2)$ (a) f(x) = $Kx^2(1-x)$ for $0 < x < 1$, (b) K/x^{a+1}, a>1, x>1, (c) f(x) = K (X^2+1) for x∈ {−2, −1, 0, 1, 2}.

Solution Integrate over the range to get $K[x^3/3 - x^4/4]_0^1 = K/12$. As the total area is 1, we get K = 12. E(X) = $12\int_0^1 x^3(1-x)dx$ = 12[1/4−1/5] = 12/20 = 3/5. $E(X^2) = 12\int_0^1 x^4(1 - x)dx$ = 12[1/5−1/6] = 12/30 = 2/5. For part (b) note that x must be greater than 1. Integrate K/x^{a+1}, to get $K(x^{-a}/ - a|_1^\infty) = K/a$ which gives K = a as the total area must be 1. E(X) = $\int_1^\infty a/x^a dx$ = a/(a−1). $E(X^2) = \int_1^\infty a/x^{a-1}dx$ = a/(a−2). The third case is discrete (as x takes integer values). Proceed as above to get K = 1. The moments are obtained by summation.

Promotional coupons in cereals pack

A cereals manufacturer offers a promotional coupon with a new brand of cereals pack. Two types of coupons (that carry either 1 point or 2 points) are printed, and exactly one of them is put in each pack. Probability that a customer will find a 1-point coupon is p, and a 2 points coupon is q = 1−p. If a customer purchases n packs of the cereal, what is the expected number of points earned?

Solution As each pack contains a coupon, the minimum score is n and the maximum score is 2n. These two scores can happen in one way each with respective probabilities p^n and q^n. The customer can score (n + 1) points in n ways (exactly one cereal pack contains a 2-point coupon, and the rest (n − 1) packs contain 1-point coupons) with probability $p^{n-1}q$, so that the probability of score n + 1 is $\binom{n}{1}p^{n-1}q$, and so on. Hence the expected score is found by summing the points earned multiplied by the corresponding probabilities as E(X) = $(n*p^n +$ $(n+1)\binom{n}{1}p^{n-1}q+\cdots +2n*q^n)$. Separate the first term from each, and use binomial expansion for $(p+q)^n = 1$ to simplify the above as n+$\binom{n}{1}p^{n-1}q + 2\binom{n}{2}p^{n-2}q^2 +\cdots + nq^n$. The "n+" term here is a guarantee that the minimum score is n. Using the identity i*$\binom{n}{i} = n * \binom{n-1}{i-1}$ this can be further simplified as n + nq(p + q)$^{n-1}$ = n + nq = n(1 + q). The maximum score of 2n occurs when q→1. If an equal number of coupons of each type are printed, q = 1/2, so that the expected score is 3n/2.

Coin tossing game

Consider a simple game in which a fair coin is tossed. You win $100 if the Head turns up. If it is a Tail that turns up, you lose $90. What is your expected loss or gain in one toss? What

Table 1.2 Number of chicken hatched in 10 d

Eggs	42	50	49	50	45	47	48	49	50	48
Chicken	37	43	47	44	40	45	44	42	48	45

is the expected value in n (>2) tosses? Does the expected value converge when a sufficiently large number of trials are conducted?

Solution As the coin is fair, Pr(Head) = Pr(Tail) = 1/2. Thus the expected value in one toss = (1/2)*100 − (1/2)*90 = 50–45 = 5, which is a gain. If this game is repeated n times, our expected gain is 5*n. This expected value is divergent as n → ∞.

The expected value can also be a negative number. In the above example, if the winning amount is 90 and losing amount is 100, the expected value is negative. Another example is given below:

Roll of a die

Consider the experiment of rolling a fair die once. If the number that comes on top is an even integer, you win c units of money. If the number on top is odd, you lose 2*c units of money. What is your expected gain or loss?

Solution If the number on top is 2, 4 or 6, the winning amount is c. If it is 1, 3, or 5, the losing amount is 2c. As the die is fair, each of them has equal probability 1/6. Let X denote the loss or gain. Then $E(X) = -2c *P(X = 1) + c * P(X = 2) -2c *P(X = 3) + c * P(X = 4) -2c *P(X = 5) + c * P(X = 6) = (1/6) [-6c + 3c] = -(3/6)c = -c/2$, which is a negative number for c> 0.

Egg hatching

A farmer hatches between 40 and 50 eggs every week. Total number of chickens hatched are given in Table 1.2. Find the expected number of chickens that will be hatched next week if n eggs are kept for hatching.

Solution We assume that the eggs are uniformly sampled from a hypothetical population with a constant probability p of hatching. From the table, we get the probabilities of hatching as $p_1 = 37/42 = 0.881$, $p_2 = 43/50 = 0.86$, $p_3 = 0.9592$, $p_4 = 0.88$, $p_5 = 0.8889$, $p_6 = 0.9575$, p7 = 0.9167, p8 = 0.8571, p9 = 48/50= 0.96, p10 = 45/48 = 0.9375. The expected probability of hatching is found as

E(p) = [0.881 + 0.86 + 0.9592 + ⋯ + 0.9375]/10 = 9.09778/10 = 0.909778. If n eggs are kept for hatching, the expected number of chickens hatched is 0.909778*n. The expected values for n = 40 to n = 50 are [36.39, 37.30, 38.21, 39.12, 40.03, 40.94, 41.85, 42.76, 43.67, 44.58, 45.49.] As the answer must be a whole integer, we could round-off the above values to the nearest integer.

1.3.1   Range for Summation or Integration

We have taken the general range for X anywhere on the real line in the above definition. Range of X depends upon the RV. For example, Poisson and Geometric distributions assume integer values in 0 to ∞, whereas binomial distribution has range 0 ⋯ n. Thus the range for summation (or integration) collapses to the range of the RV or event involved.

Expected value of marks

A multiple choice exam comprises of 50 questions, each with 5 answers. All correctly marked answers score 1 mark, and all wrong answers get a penalty of 1/4 mark. What is the expected number of marks obtained by a student who guesses n questions?

Solution Assume that the student has actually obtained x correct and (n–x) wrong answers. Then the marks obtained is x*1–(n–x)*(1/4) = (5/4)*x–n/4 = (5x–n)/4 = Y (say). As X_i can take the possible values 0, 1, …, n we get the expected value as E(X) = 1*1/5 + 0*4/5 = 1/5. Now E(Y) = (5 E(X)–n)/4 = (5x(n/5)–n)/4 = 0. Hence the answer is zero. As it is an expected value, the actual score earned when all n questions are attempted could fluctuate around the expected value.

1.3.2   Expectation Using Distribution Functions

Some of the statistical distributions have simple expressions for the CDF. Examples are the exponential, logistic, and extreme value (Gumbel) distributions. The expected value could be found in terms of the distribution functions as follows:

Theorem 1.1 *If X is discrete, then E(X) = $\sum_k P(X \geq k)$.*

Proof Without loss of generality, assume that X takes the values 1, 2, ... Then

$$E(X) = \sum_k kP(X = k) = 1 * P(X = 1) + 2 * P(X = 2) + 3 * P(X = 3) + \cdots . \quad (1.2)$$

Split 2*P(X = 2) as P(X = 2) + P(X = 2) etc. and sum alike terms to get

$$E(X) = [P(X = 1) + P(X = 2) + \cdots] + [P(X = 2) + P(X = 3) + \cdots] + \text{etc.} =$$
$$P(X \geq 1) + P(X \geq 2) + \cdots = \sum_k P(X \geq k). \quad (1.3)$$

As E(X ± c) = E(X) ± c (see Sect. 1.4), the result follows for arbitrary range of X.

Closed form summation of Incomplete Beta Function

Prove that

$$\sum_{k=0}^{n} I_p(k, n - k + 1) = np, \quad \text{where} \quad I_x(a, b) = \frac{1}{B(a, b)} \int_0^x t^{a-1}(1 - t)^{b-1} dt \quad (1.4)$$

is the Incomplete Beta Function (IBF).

Solution The CDF of X~BINO(n, p) is related to IBF as,

$$\sum_{x=k}^{n} \binom{n}{x} p^x q^{n-x} = I_p(k, n - k + 1). \quad (1.5)$$

Using the above theorem, we have

$$E(X) = \sum_k P(X \geq k) = \sum_{k=1}^{n} I_p(k, n - k + 1) = np, \quad (1.6)$$

because the mean of BINO(n, p) is np and 0*Pr(X = 0) = 0. Note that the IBF is a continuous function of p in (1.6), which is being summed. The LHS gives exact result only for small n values, as error could propagate for large n values. For instance, if n = 8 and p = 0.9 Eq. (1.6) gives 7.2, which is exact. For n = 12 and p = 0.5, Eq. (1.6) gives 5.999756 whereas np = 6. Use symmetry relation to get a similar expression

$$\sum_{k=1}^{n} I_{1-p}(n - k + 1, k) = nq, \quad (1.7)$$

which is better suited for p > 0.5. Hence the closed form expression (on the RHS) in (1.6) is extremely useful to evaluate sums on the LHS, especially when n is large.

Closed form for infinite summation of IBF

Prove that

$$\sum_{j=0}^{\infty} I_q(j, k) = kq/p, \quad \text{where} \quad q = 1 - p. \tag{1.8}$$

Solution The CDF of X∼NBINO(k, p) is related to IBF as,

$$\sum_{x=0}^{c-1} \binom{x+k-1}{x} p^k q^x = I_p(k, c). \tag{1.9}$$

Upper tail probability (SF) is obtained by subtraction as $\sum_{x=c}^{\infty} \binom{x+k-1}{x} p^k q^x = 1 - I_p(k, c)$ $= I_q(c, k)$. Using the above theorem, we have

$$E(X) = \sum_k P(X \geq j) = \sum_{j=0}^{\infty} [1 - I_p(k, j)] = \sum_{j=1}^{\infty} I_q(j, k) = kq/p \tag{1.10}$$

because the mean of NBINO(k, p) is kq/p = k(1/p–1), and 0*Pr(X = 0) = 0. Although an infinite sum, rapid convergence of (1.10) occurs for p values in the right tail (>0.5, especially for p above 0.80 or equivalently q<0.20) so that it can be truncated at j = 2k, giving the LHS as $\sum_{j=1}^{2k} I_p(j, k)$ for p > 0.5. However, many more iterations are needed when p ≤ 0.20. Using symmetry relation, it can be shown that $\sum_{j=1}^{m} I_p(k, j) = m - kq/p$. Note that the value of p must be in extreme right tail (say p>0.60) or m must be large (and k preferably <m) for this to be accurate to at least one decimal place (Table 1.3).

E(X) in terms of F(x) for continuous variates

If X is a continuous RV with CDF F(x), then

$$E(X) = \int_0^{\infty} [1 - F(x)] dx - \int_{-\infty}^{0} F(x) dx. \tag{1.11}$$

Solution Consider $\int_0^{\infty} [1-F(x)] dx$. As [1–F(x)] is the survival function (right tail area), it can be written as $[1-F(x)] = \int_x^{\infty} f(t) \, dt$ [Chakraborti et al. (2019)]. Substitute in the RHS above to get $\int_0^{\infty} [1-F(x)] dx = \int_{x=0}^{\infty} \int_{t=x}^{\infty} f(t) \, dt \, dx$. Change the order of integration to get

Table 1.3 Convergence of closed form expression for $\sum_{j=0}^{\infty} I_q(j,k)$

n	p	# iterations	Infinite sum	Closed form	n	p	# iterations	Infinite sum	Closed form
6	0.20	24	19.229616	24	14	0.20	56	48.889899	56
6	0.20	80	23.998452	24	14	0.20	80	55.045349	56
6	0.40	24	8.973200	9	14	0.40	56	20.999469	21
6	0.40	80	9.000000	9	14	0.60	56	9.333333	9.333
6	0.60	24	3.999994	4	14	0.80	56	3.500000	3.5
6	0.80	24	1.500000	1.5	18	0.20	72	63.988368	72
10	0.20	40	33.935536	40	18	0.20	100	71.038481	72
10	0.20	80	39.920751	40	18	0.40	72	26.999922	27
10	0.40	40	14.996289	15	18	0.60	72	12.00000	12
10	0.60	40	6.666667	6.667	18	0.80	72	4.500000	4.5
10	0.80	40	2.500000	2.5	20	0.40	100	30.000000	30

Accurate for p > 0.50. More iterations are needed for p values less than 0.50

$\int_{t=0}^{\infty}\int_{x=0}^{t}f(t)\,dx\,dt$. As f(t) is a constant while integrating wrt x, the inner integral simplifies to $t*f(t)$. Thus the RHS becomes $\int_{t=0}^{\infty} t*f(t)\,dt$, which is the LHS. Similarly $\int_{-\infty}^{0} F(x)\,dx = -\int_{-\infty}^{0} t*f(t)\,dt$. Combine both results to get the desired answer. We can express the quantiles using the CDF as follows: The kth quantile is that value of x for which $F(x) = k/100$. In the case of quartiles, the divisor is 4 so that we get 3 quartiles that divide the total area into 25 50 and 75 points.

Theorem 1.2 *If X is non-negative, prove that $E(X) = \int_0^{\infty}[1-F(x)]\,dx$.*

Proof By definition,

$$E(X) = \int_0^{\infty} xf(x)dx = \int_0^{\infty} xdF(x) = \lim_{t\to\infty}\int_0^{t} xdF(x) =$$

$$\lim_{t\to\infty}\left[xF(x)\big|_0^t - \int_0^t F(x)\,dx \right] = \lim_{t\to\infty}\left[tF(t) - 0 - \int_0^t F(x)\,dx \right]. \qquad (1.12)$$

Letting t tend to infinity, this becomes $F(\infty)\lim_{t\to\infty}\int_0^t dx - \int_0^t F(x)\,dx$. Put $F(\infty) - 1$ to get $\lim_{t\to\infty}\int_0^t[1-F(x)]dx$. Now let t tend to infinity to get the final results as $\int_0^{\infty}[1-F(x)]dx$.

Uniform distribution

If the CDF of a <u>discrete</u> random variable is

$$F(x) = \begin{cases} 0 & \text{if } x < 0 \\ 0.25 & \text{if } 0 \leq x < 2 \\ 0.50 & \text{if } 2 \leq x < 4 \\ 0.75 & \text{if } 4 \leq x < 6 \\ 1.0 & \text{if } x \geq 6 \end{cases}$$

find the PMF and the mean.

Solution Obviously, the CDF is a step function increasing at regular interval showing that the PMF is that of a uniform distribution as $f(x) = 0.25$ for $x = 0, 2, 4, 6$ with mean 3.

Sending POP messages

A POP (Post-Office Protocol) based mail-server sends each message and then waits for an ACK from the receiver. Only after the receipt of the ACK will the mail-server send the next message in the queue. It is known that the delay in receipt of the ACK is exponentially distributed with mean half-a-second. If 3 messages, each of size 1K are to be sent, what is the expected number of seconds elapsed for successful transmission if the sending of each message is independent and each takes half-second?.

Solution There are two non-overlapping time intervals involved. A fixed preparation delay of 0.5 s and a random duration which is EXP(0.50) so that the mean is $1/0.50 = 2$. Hence sending the message and receiving ACK takes on the average 2.5 s, so that for 3 messages, total seconds elapsed is 7.5 s on the average. This could of course vary around the true value with a lower bound of 1.5 s (total fixed delay). For example, if all mails are to local users with account on the same machine, the ACK delay will be negligible.

E(X) and E(X²) of a Poisson distribution

Find $E(X)$ and $E(X^2)$ for a Poisson RV.

Solution Let X~ POI(λ). Then $E(X) = \sum_{x=0}^{\infty} x e^{-\lambda} \lambda^x / x! = e^{-\lambda} \lambda \sum_{x=1}^{\infty} \lambda^{x-1}/(x-1)! = e^{-\lambda} \lambda e^{\lambda} = \lambda$. To find $E(X^2)$, write $X^2 = X*(X-1) + X$. Then $E(X^2) = \sum_{x=0}^{\infty} x(x-1)e^{-\lambda}\lambda^x/x! + \sum_{x=0}^{\infty} x e^{-\lambda}\lambda^x/x!$. The x(x-1) in the first sum cancels out with x! in the

denominator giving (x–2)!. Thus the first term becomes $\lambda^2 e^{-\lambda} \sum_{x=2}^{\infty} \lambda^{x-2}/$
$(x-2)!$. Upon putting x–2 = y, the summation reduces to e^{λ} giving $\lambda^2 e^{-\lambda} e^{\lambda} = \lambda^2$. Substitute for the second sum from above to get $E(X^2) = \lambda^2 + \lambda$. We know that the tail probabilities of a Poisson distribution is related to the incomplete gamma function as follows:

$$F(r) = P[x > r] = \sum_{x=r+1}^{\infty} e^{-\lambda} \lambda^x / x! = \frac{1}{\Gamma(r+1)} \int_0^{\lambda} e^{-x} x^r dx = P_{r+1}(x) \qquad (1.13)$$

Substitute for P[x>r] = $P_{r+1}(x)$ using Theorem 1.1 to get E(X) = $\sum_k P(X \geq k)$ = $\sum_r P_{r+1}(x)$.

Mathematical expectation can also be defined on events associated with a RV. Consider the event Y that a Poisson RV X takes even values. The possible values of Y are 0, 2, 4 , ... , ∞. Then E(Y) = $\sum_{y\ even}^{\infty} y e^{-\lambda} \lambda^y / y!$. Put u = y/2, so that u takes all integer values starting with 0. As before E(U) = $\sum_{u=0}^{\infty} u e^{-\lambda} \lambda^u / u! = \lambda$, so that E(Y) = 2λ.

Application of Poisson distribution

The number of WhatsApp messages arriving in a cell phone between 9 AM and 5 PM is Poisson distributed with λ = 5 for a 20 min interval. What is the expected number of messages received in one hour? What is the total expected number of messages received between 1 PM and 5 PM?

Solution As the expected value of a Poisson RV is λ we expect five messages on the average in 20 min, so that on the average there will be 3 × 5 = 15 messages per hour. This does not mean that it will always be 15, but it could be any number between 0 and a maximum, and it will fluctuate around the central value 15. Similarly on the average there will be 4 × 15 = 60 messages between 1 PM and 5 PM.

Software Exceptions

A software comprises of 8 subsystems. Probability that the first 5 subsystems will throw a run-time exception (kind of error) in 8 h of use is Poisson distributed with λ = 0.03, and independently the last 3 subsystems is Poisson distributed (λ = 0.05). If the software is used for 80 h, (i) what is the expected number of exceptions? (ii) probability that no exceptions occurred.

Solution (i) Expected number of exceptions in 8 h = 0.03 + 0.05 = 0.08, so that in 80 h we expect 0.08 * 10 = 0.8 exceptions. (ii) Probability of no exceptions in 8 h = exp(–0.03) *

$\exp(-0.05) = 0.9704455 * 0.951229 = 0.923116$, so that in 80 (=8 * 10) hours probability of no exceptions = 0.44933.

Derangement

Suppose n letters are to be sent in n envelops. If the letters and envelops are shuffled and each letter is randomly put in an envelope, find the expected number of matches (letters that get into correct envelops).

Solution Let X denote the event that a letter is assigned to its correct envelope. The number of letters that go into correct envelopes can take the values x = 0, 1, 2, ..., n. Then x = 0 indicates that none of the letters go into correct envelope. This can happen in D_n ways, so that the probability of this event happening is $1/D_n$ where D_n denotes derangement of n objects. One letter can go into the correct envelope when (n–1) letters do not get their correct envelope. This can happen in $\binom{n}{1}D_{n-1} = nD_{n-1}$ ways with probability $1/[nD_{n-1}]$. Continuing similarly we find that k letters go into correct envelope when (n–k) remaining letters are deranged in $\binom{n}{k}D_{n-k}$ ways. All n letters go into their respective envelope in just one way. If (n–1) letters go into their correct envelope, the remaining one must automatically go into the correct envelope. Hence there exist only one possibility in these two cases. The expected number of letters that go into the correct envelope can be found using $\sum_{x=0}^{n} x\binom{n}{x}D_{n-x}$. Write $\binom{n}{x} = (n/x)\binom{n-1}{x-1}$ to get $n\sum_{x=1}^{n}\binom{n-1}{x-1}D_{n-x}$. This can be evaluated using generating functions [Chattamvelli and Shanmugam (2020)].

Find unknowns in an expectation

If X is a continuous RV, and d is a constant, find the unknowns a, b, c such that $E[X-d|X>d] = \int_a^b c[1-F(x)]dx$ where b>a.

Solution Consider the probability Y = P[X–d|X > d]. Then P[Y ≤ y] = P[X-d|X>d ≤ y]. This probability is the same as P[X–d ≤ y|X > d]. As X is continuous, this can be written as P[d < X < y + d|X > d]. Using the conditional probability formula P[A|B] = P[A∩B]/P[B] this becomes P[d < X < y + d]/P[X > d]. The numerator can be written in terms of unconditional CDF of X as $F_X(y + d)-F_X(d)$ if d is positive, and $F_X(d)-F_X(y-d)$ if d is negative. The denominator being the survival function can be written as $1-F_X(d)$. Now apply Theorem 1.2 on the expectation of Y as $E[Y] = \int_0^\infty [1-G_Y(y)]$ dy. Substitute for $G_Y(y)$ from the above to get $E[Y] = \int_0^\infty [1-[F_X(y + d)-F_X(d)]/[1-F_X(d)]]$ dy. This simplifies to $E[Y] = \int_0^\infty [1-[F_X(y + d)]/[1-F_X(d)]dy$. Apply a change of variable transformation x = y + d to get $E[Y] = \int_d^\infty [1-[F_X(y)]/[1-F_X(d)]dx$. From this we see that a = d, b = ∞ and c = $1/[1-F_X(d)]$.

Variance in terms of F(X)

If X is continuous RV, the variance of X can be expressed as $\sigma_x^2 = \int_0^\infty 2x[1 - F_X(x) + F_X(-x)]dx - \mu_X^2$.

Solution As $\text{Var}(X) = E[X^2] - (E[X])^2 = E[X^2] - \mu_x^2$. Substitute for $= E[X^2]$ to get the result.

Expectation of Integer part of Exponential

If $X \sim \text{EXP}(\lambda)$ find $E(\lfloor X \rfloor)$ where $\lfloor X \rfloor$ denotes the integer part of X.

Solution The integer part of X has a geometric distribution with $q = \exp(-\lambda)$ [Chattamvelli and Shanmugam (2021)]. Using the above method, the problem reduces to finding the mean of GEO(p) where $p = 1-q = [1 - \exp(-\lambda)]$. Hence $E(Y) = E(\lfloor X \rfloor)$ is $\exp(-\lambda)/[1 - \exp(-\lambda)]$.

1.4 Expectation of Functions of RVs

Simple mathematical functions of RVs appear in several practical applications in engineering and applied sciences. A possible method is to first find the distribution of these functions and then find its expected value. Let g(x) be a real function without discontinuities. First find the distribution of Y = g(X) using a technique discussed in next Chapter, and find E[Y] of this distribution. As an example, let $X \sim N(0, 1)$, and $Y = X^2$. We wish to find $\mu_2' = E[X^2] = \int_{-\infty}^\infty x^2 \phi(x)dx$, where $\phi()$ is the standard normal PDF. We know that $Y = X^2$ has a central χ^2 distribution with 1 degree of freedom, whose expected value is 1. Hence $E[X^2]$ is also 1. This technique may not always work. In the above example, if we wanted $E[X^2 - 2X + 3]$, we need to resort to the following approach because $X^2 - 2X + 3$ does not have a simple distribution. The following section gives us a simple method to find expected values of functions of RVs without either deriving their distributions or knowing about the exact distributions.

Expected value of $\exp(-\lambda X)$

If X is uniformly distributed in [a, b] find the expected value of $\exp(-\lambda X)$.

Solution We know f(x) = 1/(b–a), so that $E(\exp(-\lambda x)) = 1/(b–a)\int_a^b \exp(-\lambda x)dx = 1/(b–a)\exp(-\lambda x)/ - \lambda|_a^b = 1/(b–a)[\exp(-b\lambda) - \exp(-a\lambda)]$.

1.4.1   Properties of Expectations

Let X and Y be any two RVs, discrete or continuous, univariate or multivariate. In the following discussion it is assumed that E(X) and E(Y) exist (they are finite).

Theorem 1.3 *The expected value of a constant is constant.*

Proof The proof follows trivially because the constant can be taken outside the summation (discrete case) or integration (continuous case) and what remains is either the summation or integration of probabilities that evaluates to a 1.0. Symbolically, E(c) = c. Here c is a scalar constant for univariate distributions, and a constant vector for multivariate distributions. Symbolically, $E(c) = \sum_k c p_{x=k} = c \sum_k p_{x=k} = c$, for the discrete case. If X is continuous, $E(c) = \int_x c p(x)dx = c \int_x p(x)dx = c$.

Theorem 1.4 *The expected value of linear function c*X is c times the expected value of X, where c is a nonzero constant and the expected value exists.*

Proof As above, the constant can be taken outside the summation (discrete case) or integration (continuous case) and what remains is either the summation or integration of X that evaluates to E(X). Applying the multiplier c gives the result as c*E(X).

Theorem 1.5 *Prove that expected value of linear combination E(a*X + b) = a* E(X) + b for any RV X, and nonzero constant a.*

Proof Let X be discrete, and take values $x_1, x_2, \ldots, x_\infty$. From the definition of expected values, $E(aX + b) = \sum_k (ax_k + b) p_{x_k} = a\sum_k x_k p_{x_k} + b\sum_k p_{x_k} = a\ E(X) + b$ because $\sum_k p_{x_k} = 1$. If X is continuous, $E(aX+b) = \int (ax + b)p(x)dx = a\int xp(x)dx + b\int p(x)dx = a\ E(X) + b$. We have not made any assumption on the distribution of the RV X in this theorem, but only the existence of the first moment.

Corollary 1.1 *E(c–X) = c – E(X). This follows by writing (c–X) as (–1*X + c) and applying the above theorem with a = –1, and b = c. Putting c = 0, we get E(–X) = –E(X).*

E(n–X) of a Binomial

If X has a binomial distribution with parameters n and p (BINO(n, p)), find E(n–X), Var(n–X).

Solution Write n–X as Y = (–1)*X + n, and apply Theorem 1.5 to get E(Y) = (–1)*E(X) + E(n). Substitute E(X) = np, and use E(c) = c to get E(Y) = –np + n = n(1–p) = nq. Var(Y) = Var(n–X) = $(-1)^2$ Var(X) = Var(X) = npq, using Var(aX + b) = a^2 Var(X).

Theorem 1.6 *If X and Y are two RVs, E(X ± Y) = E(X) ± E(Y) = E(Y ± X), and E(aX ± bY) = a E(X) ± b E(Y).*

Proof The first result follows trivially by distributing the summation or integration over the individual components (connected by + or –). The second result can be proved using the fact that E(cX) = c E(X), twice. This is called the linearity property of expectation.

1.4.2   Expected Value of Independent RVs

For independent random variables Pr(XY) = Pr(X) * Pr(Y).

This result is defined in terms of probabilities. As expected values can be considered as functions of RVs with probabilities as weights, we could get analogous results in terms of expected values. Two outcomes are independent if knowing the outcome of one does not change the probabilities of the outcomes of the other. When two events are independent, we find the probability of both events happening by multiplying their individual probabilities.

Definition 1.9 If X and Y are two *independent* RVs, then E(XY) = E(X) * E(Y).
Let X and Y be discrete. Then $E(XY) = \sum_k (x_k y_k) p_{x_k, y_k} = \sum_k x_k y_k p_{x_k} p_{y_k}$ using the above theorem on the independence of probabilities. Pairing x_k with p_{x_k}, this becomes $\sum_k (x_k p_{x_k})(y_k p_{y_k})$. Hence if X and Y are independent, then E[XY] = E[X]E[Y].

Theorem 1.7 *If X and Y are two independent RVs, then Pr(X ≤ x, Y ≤ y) = Pr(X ≤ x) * Pr(Y ≤ y).*

Proof Let both be discrete. Write the LHS as

$$\Pr[X \le c, Y \le d] = \sum_{x \le c} \sum_{y \le d} p(x, y) = \sum_{x \le c} \sum_{y \le d} p(x)p(y) \qquad (1.14)$$

Using the properties of summation mentioned in this chapter, split this into two products as

$$\Pr[X \le c, Y \le d] = \sum_{x \le c} p(x) \sum_{y \le d} p(y) = \Pr[X \le c] * \Pr[Y \le d] \qquad (1.15)$$

due to the independence of X and Y.

Function of independent random variables

If X and Y are independent RVs with E(X) = –3 and E(Y) = 5, find E(2X–3)(Y + 5).

Solution E(2X–3)(Y + 5) = E[2XY + 10X–3Y–15]. As X and Y are independent, E(XY)= E(X)* E(Y). Substitute given values to get E(2X–3)(Y + 5) = 2*(–3)*5 + 10*(–3)–3*5–15 = –90.

Theorem 1.8 *Expected value of a linear combination of scaled functions is the linear combination of expected value of the functions with respective scaling factors. Symbolically,* $E(c_1 g_1(x) + c_2 g_2(x) + \cdots +) = c_1 E(g_1(x)) + c_2 E(g_2(x)) + \cdots +,$ *where the constants* c_i*'s are any real numbers.*

Proof This result follows trivially using theorem and corollaries above. Putting $c_1 = +1$ and $c_2 = -1$ we get $E([g_1(x) - g_2(x)]) = E(g_1(x)) - E(g_2(x))$.

Definition 1.10 If X and Y are two *independent* RVs, and g(x), h(y) are everywhere continuous functions of X and Y defined on the range of X and Y, then E(g(X)*h(Y)) = E(g(X)) * E(h(Y)) if the expectations on the RHS exist.

The proof follows exactly as done above using summation in the discrete case and integration in the continuous case. A direct extension of this result is that $E(\prod_{k=1}^{n} a_k X_k) = \prod_{k=1}^{n} a_k E(\prod_{k=1}^{n} X_k)$. This simplifies to $\prod_{k=1}^{n} (a_k E[X_k])$ when they are independent.

Find normalising constant

The PMF of a discrete RV is given by $f(x) = cx^2$ for x={1, 2, 3}. Find the mean and variance.

Solution As the total probability is 1, c[1+4+9] = 1 from which c = 1/14. E(X) = $1/14\sum_x x^3$ = [1 + 8 + 27]/14 = 36/14 = 18/7 = 2.57142857. E(X^2) = $1/14\sum_x x^4$ = [1+16+81]/14 = 98/14 = 7. Hence variance = 7–6.612244897 = 0.387755.

Find normalising constant

If f(x) = K x(x + 1) for x = 1, 2, 3, 4; find E[X] and P[X ≥ 2].

Solution As total probability is 1, we get K[2 + 6 + 12 + 20] = 1 which gives K = 1/40. E(X) = $\sum_x x^2(x + 1)$ = [2 + 12 + 36 + 80]/40 = 130/40 = 3.25. To find P[X ≥ 2] we use complement rule to get 1–P[X = 1] = 1–2/40 = 0.95

Correlation between (i) (X + c and Y + d), (ii) (cX, dY)

Find the correlation between X + c and Y + d where c and d are constants.

Solution Let ρ denotes $\text{Corr}(X, Y) = \text{Cov}(X, Y)/[\sqrt{Var(X)}\sqrt{Var(Y)}]$. As covariance and variance are both unchanged due to a change of origin transformation, $\text{Corr}(X + c, Y + d)$ is the same as $\text{Corr}(X, Y)$. In case (ii), the numerator expression is $\text{Cov}(cX, dY) = cd \, \text{Cov}(X, Y)$. As $\text{Var}(cX) = c^2 \, \text{Var}(X)$ and $\text{Var}(dY) = d^2 \, \text{Var}(Y)$, the constants cancel out from numerator and denominator. Thus we get $\text{Corr}(cX, dY) = \text{sign}(cd) \, \text{Corr}(X, Y)$. Symbolically, $\rho_{cX,dY} = \text{sign}(cd) * \rho_{X,Y}$.

$E(X^2)$ using $E(X(X-1))$

Prove that $E[x^2]$ can be found easily using $E[x(x-1)]$ and $E[x]$ when the denominator of the RV involves factorials. Find a similar method to find $E[x^3]$. Use this technique to find the expected values of a Poisson RV.

Solution Write $x^2 = x(x-1) + x$. Use the above result to break the expected value of RHS into two terms to get $E[x^2] = E[x(x-1)] + E[x]$. Write $x^3 = x(x-1)(x-2) + 3x(x-1) + x$. Take expectation to get $E(x^3) = E[x(x-1)(x-2)] + 3E[x(x-1)] + E[x]$. For a Poisson RV, $E[x(x-1)(x-2)] = \sum_{x=0}^{\infty} x(x-1)(x-2) \exp(-\lambda)\lambda^x/x! = \sum_{x=3}^{\infty} \exp(-\lambda)\lambda^x/(x-3)! = \lambda^3$. Similarly, $E[x(x-1)] = \lambda^2$. Put these values in the above expression to get $E(x^3) = \lambda^3 + 3\lambda^2 + \lambda = \lambda(\lambda^2 + 3\lambda + 1) = \lambda[(\lambda + 1)^2 + \lambda]$.

Sum of IID Bernoulli RVs

If $X \sim \text{BER}(p_1)$, and $Y \sim \text{BER}(p_2)$ are IID, then $E(X + Y) = p_1 + p_2$.

Solution As X and Y are compatible, we apply the above theorem to get the result. This theorem can be extended to any number of RVs.

Theorem 1.9 *If $X_1, X_2, \ldots, X_m$ are BER(p_i), then $E(X_1 + X_2 + \cdots + X_m) = p_1 + p_2 + \cdots + p_m = \sum_{i=1}^{m} p_i$.*

Proof The proof follows easily by taking the expectation term by term. If the probability of success is equal (the same probability p for each of them) then $E(X_1 + X_2 + \cdots + X_m)$ = mp.

Theorem 1.10 *If $X_1, X_2, \ldots, X_n$ are RVs, each with mean μ, the mean of $\overline{X}_n = (X_1 + X_2 + \cdots + X_n)/n$ is also μ.*

Proof Take 1/n as a constant on the RHS and apply the above theorem to get $E[\overline{X}_n] = (1/n)*$ $(E[X_1] + E[X_2] + \cdots + E[X_n] = (1/n)*n\mu = \mu$ if they are independent.

Theorem 1.11 *If $X \le Y$, then $E[X] \le E[Y]$.*

Proof For simplicity assume that X and Y have the same range. Consider Z = Y–X. As $X \le Y$, Z is always positive. Hence $E[Z] \ge 0$. This means that $E(Y–X) \ge 0$, or equivalently $E[Y] \ge E[X]$.

As noted above, the sum on the left makes sense only when the RVs are compatible. Any of the "+" can also be replaced by a "−" with the corresponding sign changed on the RHS accordingly.

Sum of numbers drawn

Suppose an urn contains n indistinguishable coins numbered 1 to n. If r coins are drawn without replacement, what is the expected sum of the numbers on the coins?

Solution There are $\binom{n}{r}$ ways to choose r numbers out of n. Let X denote the sum of the numbers on the coin. Obviously the minimum sum is L = (1 + 2 + 3 + $\cdots$ + r) = r(r + 1)/2 and maximum sum is U = (n − r + 1) + (n − r + 2) + $\cdots$ + n = r(2n − r + 1)/2. These two extreme cases can happen just one way, as also the sums (r(r + 1)/2) + 1 = L + 1 and (r(2n − r + 1)/2)–1 = U–1 so that the probability for each of them is $1/\binom{n}{r}$. For all other intermediate values there exist 2 or more ways. Let k be an intermediate sum other than the 4 mentioned above. Then the number of ways to get a sum of k using r numbers is given by Stirling number of second kind S(k, n) so that the corresponding probability is $S(k, n)/\binom{n}{r}$. The expected value is then found using $\sum_{x=L}^{U} x * S(x, n)/\binom{n}{r}$ where S(L, n) = S(L + 1, n) = S(U–1, n) = S(U, n) = 1.

Expected value of functions of Binomial

If X has a binomial distribution with parameters n and p (BINO(n, p)), then (i) E(X/n) = p, and (ii) $E(n–X)^2 = nq[n–p(n–1)] = nq + n(n–1)q^2$.

Solution The first result follows trivially from the above by replacing c with 1/n, and taking 1/n as a constant outside the expectation operator. For (ii) expand the quadratic as $E(n − X)^2$

$= E(n^2 - 2nX + X^2)$. Take term by term expectation to get $E(n^2) - 2nE(X) + E(X^2))$. Now apply above theorems to get $n^2 - 2n*np + (np + n(n-1)p^2)$. This simplifies to $nq[n-p(n-1)]$. Write p as $1-q$ so that $[n-p(n-1)] = n-(1-q)(n-1)$. The n cancels out giving $1 + q(n-1)$. Substitute in the above to get $E(n - X)^2 = nq[1 + (n-1)q] = nq + n(n-1)q^2$. This result can also be obtained directly from the observation that $n-X \sim$ BINO(n, q) so that its second moment is $nq + n(n-1)q^2$.

Expectation of minimum of GEO(p).

If X and Y are IID GEO(p) and GEO(q) distributions, find E[min(X, Y)] and E[max(X, Y)].

Solution   It is shown in Chap. 2 (2.9) that min(X, Y) $\sim$ GEO(1−(1−p)(1−q)) which is the same as GEO(p + q − pq). From this E[min(X, Y)] follows directly as $1/(p+q-pq)$ ([Chattamvelli and Shanmugam (2021)]). Now use E[max(X, Y)] = E(X) + E(Y) − E[min(X, Y)] = $1/p + 1/q - 1/(p + q - pq)$.

Expectation Inequality

Prove that $[E(XY)]^2 \leq E(X)^2 * E(Y)^2$ if X and Y are real-valued.

Solution   Consider the expression $(aX+Y/a)^2 = a^2 X^2 + Y^2/a^2 + 2*a*(1/a)*XY = a^2 X^2 + Y^2/a^2 + 2*XY$. As the LHS being a square is always non-negative, this can be written as $\pm 2XY \leq a^2 X^2 + Y^2/a^2$. Take expectation of both sides to get $\pm 2E[XY] \leq a^2 E[X^2] + (1/a^2)E[Y^2]$. If $E[X^2] > 0$, take $a^2 = \sqrt{E[Y^2]}/\sqrt{E[X^2]}$ to get the RHS as $2*\sqrt{E[X^2]}\sqrt{E[Y^2]}$. Cancel out 2 from both sides to get the result.

Expected heads in four flips of a coin

What is the expected number of Heads in four flips of a coin?

Solution   Let X be the number of heads. It can take values 0, 1, 2, 3, 4 with probabilities $P(X - 0) = P(\text{"TTTT"}) = q^4$, $P(X = 1) = \binom{4}{1}pq^3 = 4pq^3$, $P(X - 2) - \binom{4}{2}p^2q^2 = 6p^2q^2$, $P(X = 3) = \binom{4}{3}p^3q = 4p^3q$ and $P(X = 4) = P(\text{"HHHH"}) = p^4$. Hence $E(X) = 0*q^4 + 1*4pq^3 + 2*6p^2q^2 + 3*4p^3q + 4*p^4 = 4[pq^3 + 3p^2q^2 + 3p^3q + p^4]$. As the coin is fair, p = 1/2. Thus the required expected number is $\frac{4}{2^4}[1 + 3 + 3 + 1] = 8/4 = 2 = np$.

Corollary 1.2 $|E[X]| \leq \sqrt{E[X^2]}$.
   This follows easily by taking Y=X.

Expectation Limit

Prove that $\lim_{\lambda \to \infty} \sum_{k=0}^{\lfloor \lambda \rfloor} \exp(-\lambda)\lambda^k/k! = 1/2$.

Solution The expression inside the summation can be considered as the PMF of Poisson distribution $X_\lambda \sim \text{POIS}(\lambda)$. As the mean and variance are both λ, the standardized variable $Z_\lambda = (X_\lambda - \lambda)/\sqrt{\lambda}$ tends to standard normal distribution using central limit theorem, so that $\Pr[Z_\lambda \leq 0] = \lim_{\lambda \to \infty} \sum_{k=0}^{\lfloor \lambda \rfloor} \exp(-\lambda)\lambda^k/k! \to \Phi(0) = 0.50$.

1.4.3   Expectation of Continuous Functions

Some applications involve functions of RVs. Examples are fractional powers of X, integer powers of X, exponential, logarithmic and trigonometric functions and other transcendental functions.

Expected value of exp(λX)

If X is BINO(n, p) find expected value of $\exp(\lambda X)$ where λ is a nonzero constant.

Solution By definition $E[\exp(\lambda X)] = \sum_{k=0}^{n} \exp(\lambda k) \binom{n}{k} p^k q^{n-k}$. Combine $\exp(\lambda k)$ with p^k and write this as $\sum_{k=0}^{n} \binom{n}{k} (p \exp(\lambda))^k q^{n-k}$. This simplifies to $(q + pe^\lambda)^n$.

Corollary 1.3 *If g(x) is undefined for at least one value of X, then E(g(x)) does not exist. For instance the first inverse moment E(1/X) is undefined for all RVs that assume a nonzero value for x = 0 (i.e.: f(x = 0) ≠0).*

$E(1/X) \geq 1/E(X)$

Prove that $E(1/X) \geq 1/E(X)$ for positively defined RVs X.

Solution Let μ be the mean of X. Then $E[(X-\mu)(1/X)-1/\mu] = E[(X-\mu)(\mu-X)/\mu X = -E[(X-\mu)^2/(\mu X) \leq 0$.

Definition 1.11 If the function g(x) is everywhere continuous in the range of X, then

$$E(g(x)) = \begin{cases} \sum_{x=-\infty}^{\infty} g(x)f(x) & \text{if X is discrete;} \\ \int_{-\infty}^{\infty} g(x)f(x)dx & \text{if X is continuous.} \end{cases}$$

Theorem 1.12 $E(g(X, Y)) = \underset{over\ x}{E} \left[\underset{over\ y}{E} (g(X, Y)|X) \right].$

Expected value of exponential function

If X is EXP(λ), find $E[e^{-X/2}]$.

Solution As X is EXP(λ), f(x;λ)=(λ) exp($-\lambda x$). Hence $E[e^{-x/2}] = \int_0^\infty e^{-x/2}(\lambda) \exp(-\lambda x)$ dx. Take the constant outside the integral and combine the exponents to get $E[e^{-x/2}] = (\lambda) \int_0^\infty \exp(-x(\frac{1}{2} + \lambda))dx$. This evaluates to $\lambda/[1/2 + \lambda]$. Had we taken the PDF as f(x; λ) $= (1/\lambda) \exp(-x/\lambda)$ we would have got $2/(\lambda + 2)$.

$E((-1)^X)$ of a Poisson Distribution

If X is POIS(λ), find $E[(-1)^x]$.

Solution This follows easily as $E[(-1)^x] = \sum_{x=0}^{\infty}(-1)^x \exp(-\lambda)\lambda^x/x!$. Using the above theorem, we take the constant outside the summation to get

$$E[(-1)^x] = \exp(-\lambda) \sum_{x=0}^{\infty} (-\lambda)^x/x! = \exp(-\lambda)\exp(-\lambda) = \exp(-2\lambda). \qquad (1.16)$$

Moments of geometric distribution $q^{x/2}p$

Find the mean and variance of a distribution defined as

$$f(x; p) = \begin{cases} q^{x/2}p \text{ if} & \text{x ranges from } 0, 2, 4, 6, \ldots, \infty \\ 0 & \text{elsewhere.} \end{cases}$$

Solution Put Y = X/2 to get the standard form. Take expectation of both sides to get E(Y) = E(X)/2 = q/p, so that E(X) = 2q/p. Similarly, V(Y) = V(X)/4 = (q/p^2)/4 = q/(4p^2).

1.4.4   Variance as Expected Value

The population variance is a measure of the spread in a population. This is captured in a single parameter for location-and-scale (LAS) distributions like the normal, Laplace, general Cauchy, Logistic, Gumbel and some other distributions. It is a function of a single parameter for Poisson, T, Maxwell and Rayleigh distributions, and is a function of two or more parameters for some others.

Definition 1.12   Variance of a RV is $\sigma_x^2 = E[(X - E(X)]^2 = E(X^2) - E(X)^2$. Here E(X) is the population mean, which we denote by μ_x. If $\mu_x = 0$, the population variance takes the simple form $\sigma_x^2 = E(X^2)$. The above can be expressed as $E[(X - E(X)]^2 = \sum_k (x_k - \mu)^2 p_X(x_k)$ when X is discrete, and $\int_x (x - \mu)^2 f(x) \, dx = \int_x (x - \mu)^2 dF(x)$ when X is continuous.

Properties of Variance

1. Var(X) = $E[X^2]$–$E[X]^2$.
    The proof follows trivially by expanding $(x - \mu)^2 = x^2 - 2\mu x + \mu^2$, then taking expectation term by term, and using $E[X] = \mu$ in the middle term (see Table 1.6).
2. The variance of independent RVs are additive. Symbolically V(X + Y) = V(X) + V(Y). This is known as the additivity property, which is valid for any number of independent RVs. We prove it for two variates X and Y. By definition, Var(X + Y) = E[(X + Y) − E[X + Y]]^2. Use E[X + Y] = E[X] + E[Y] in the inner expectation, and combine with (X + Y) to get RHS as E[(X–E[X]) + (Y–E[Y])]^2. Expand as a quadratic to get E[(X–E[X])^2 + (Y–E[Y])^2 + 2(X–E[X])(Y–E[Y])]. Now take term by term expectation and use E(X–E[X])(Y–E[Y]) = 0 (as X and Y are independent, E(XY) = E(X)*E(Y) so that E(X–E[X])(Y–E[Y]) = 0) to get the result.
3. Var(c∗X) = c^2∗Var(X).
    By definition Var(c∗X) = E[cX–E(cX)]^2. As c is a constant, it can be taken outside the expectation to get c^2∗E[X–E(X)]^2.
4. Var(c∗X ± b) = c^2∗Var(X).
    By definition Var(c∗X + b) = E[cX + b – E(cX + b)]^2. The +b and −b cancels out giving

$E[cX-E(cX)]^2 = c^2*\mathrm{Var}(X)$. Replace b by $-b$ to get a similar result. The above two results allow us to find the variance of any linear combination of RVs by finding the $\mathrm{Var}(X)$ just once, and doing simple arithmetic with the constants to get the desired result.

5. $\mathrm{Var}(X \pm b) = \mathrm{Var}(X)$.

By definition $\mathrm{Var}(X + b) = E[X + b - E(X + b)]^2$.

The $+b$ and $-b$ cancels out giving $E[X-E(X)]^2 = \mathrm{Var}(X)$. Replace b by $-b$ to get a similar result. This result shows that a change of origin transformation does not affect the variance.

6. $\mathrm{VAR}(\sum_{j=1}^{m} X_j) = \sum_{j=1}^{m} \mathrm{VAR}(X_j)$ if X'_js are independent.

This can be proved by induction using the above result 1.

Theorem 1.13 *If $X_1, X_2, \ldots, X_n$ are RVs, each of which are pair-wise uncorrelated with the same mean μ and variance σ^2, then the variance of $\overline{X}_n = \sigma^2/n$.*

Proof Consider $\mathrm{Var}(\overline{X}_n) = \mathrm{Var}((X_1 + X_2 + \cdots + X_n)/n) = 1/n^2*[\mathrm{Var}(X_1 + X_2 + \cdots + X_n)] = 1/n^2*n\sigma^2 = \sigma^2/n$.

Variance of Y = n−X of binomial distribution

If X has a binomial distribution with parameters n and p, derive the variance of $Y = n - X$.

Solution This is already derived in Sect. 1.32. Here we use the above property to derive it. $\mathrm{Var}(Y) = \mathrm{Var}(n-X) = \mathrm{Var}(n) + (-1)^2\mathrm{Var}(X)$. As the variance of a constant is zero, the RHS simplifies to $\mathrm{Var}(X)$. Hence $V(X) = \mathrm{Var}(Y) = npq$. This is obtainable directly because $n-X \sim \mathrm{BINO}(n, q)$.

Variance of points earned in Cereals coupon

In the cereals coupon Example 1.3 find the variance on the number of points earned.

Solution Let X_i denote the event associated with ith packet. Then X_i takes the value 1 with probability p and 2 with probability $1-p$ so that the expected value is $1.p + 2.(1 - p) = 2 - p$. Write this as $1 + (1 - p) = 1 + q$. If $X \geq 1$ packets are bought, $E(X) = X_1 + X_2 + \cdots + X_n = n(1 + q)$. $E(X^2) = 1^2 * p + 2^2 * q = p + 4q = 1 + 3q$ using $p + q = 1$. From this $V(X_i) = E(X_i^2) - E(X_i)^2 = 1 + 3q - (1 + q)^2 = 3q - 2q - q^2 = q - q^2 = pq$. $V(X) = V(X_1 + X_2 + \cdots + X_n) = npq$.

1.4.5   Covariance as Expected Value

Covariance is a non-standardized measure of the dependency between the variables involved. We denote it by Cov(X, Y). The order of the variables X and Y is unimportant, as it is symmetric in the variables involved.

Definition 1.13 Covariance of two RVs X and Y is Cov(X, Y) = E(XY) – E(X)E(Y) = E(X–E[X])(Y–E[Y]) = E(Y–E[Y])(X–E[X]).

Properties of Covariance

Covariance satisfies several interesting properties listed below. It is assumed that both X and Y are quantitative.

1. The covariance of two independent RVs is zero
    This follows from the above definition because E(XY) = E(X)*E(Y) when X and Y are independent.
2. The covariance of a random variable with an independent linear combination is additive. Symbolically, Cov(X + Y, Z) = Cov(X, Z) + Cov(Y, Z). By definition LHS =

$$Cov(X + Y, Z) = E(X + Y)Z - E(X + Y)E(Z) \qquad (1.17)$$

    As Z is independent of X + Y, E(X + Y)Z = E(XZ) + E(YZ). Also, E(X + Y) = E(X) + E(Y). Substitute in RHS to get E(XZ) + E(YZ)–[E(X) + E(Y)] E(Z). Rewrite this as [E(XZ)–E(X)E(Z)] + [E(YZ)–E(Y)E(Z)] = Cov(X, Z) + Cov(Y, Z). Similar results can be derived for Cov(X–Y, Z) = Cov(X, Z)–Cov(Y, Z) (if Z is independent of X–Y), Cov(X, Y + Z) = Cov(X, Y) + Cov(X, Z) (if X is independent of Y + Z), and Cov(X, Y–Z) = Cov(X, Y)–Cov(X, Z) (if X is independent of Y–Z).
3. Cov(X, Y) = E(XY) when either or both of E(X) or E(Y) = 0
    This follows easily from the definition Cov(X, Y) = E(XY)–E(X)*E(Y).
4. Cov(X, Y) = $-\mu_x * \mu_y$ when X and Y are orthogonal
    This follows from the fact that E(XY) = 0 under orthogonality.
5. Cov(a*X, b*Y ± c*Z) = ab*Cov(X, Y) ± ac*Cov(X, Z).
    LHS = E[a*X–aE(X)][(b*Y ± c*Z) – (b*E(Y) ± c*E(Z))]. Split the second expression into two and combine with the first expression to get E[a*X–aE(X)][b*Y–E(b*Y)] ± E[a*X–aE(X)][c*Z–E(c*Z)]. Take out a, b as constants from the first; and c, d as constants from the second expression to get the result. If c = 0, we get Cov(a*X, b*Y) = ab*Cov(X, Y).
6. Cov((X–a)/c, (Y–b)/d) = Cov(X, Y)/(cd)
    As the change of origin transformation does not affect the covariance, the LHS is equal to Cov(X/c, Y/d). Now apply above result with a = 1/c, b = 1/d to get the result.

7. If $U = a*X + b*Y$ and $V = c*X + d*Y$, where a, b, c, d are nonzero constants, then $\text{Cov}(U, V) = a*c\sigma_x^2 + b*d\ \sigma_y^2 + (a*d+b*c)\ \text{Cov}(X, Y)$. This result allows us to find the covariance of two arbitrary linear combinations. The proof follows exactly in the same way as above.

8. If $U = \cos(\theta) * X - \sin(\theta)*Y$ and $V = \sin(\theta) * X + \cos(\theta)*Y$, then $\text{Cov}(U, V) = \cos(\theta)*\sin(\theta)[\sigma_x^2 - \sigma_y^2] + [\cos^2(\theta) - \sin^2(\theta)]\ \text{Cov}(X, Y)$. As θ is a constant, take a $= \cos(\theta)$, b $= -\sin\theta$, etc. and apply the above theorem.

9. $\text{COV}(\sum_{j=1}^{m} X_j, \sum_{k=1}^{n} Y_k) = \sum_{j=1}^{m} \sum_{k=1}^{n} \text{COV}(X_j, Y_k)$ if the X's and Y's are independent. This allows us to find the covariance of sums of RVs.

10. $\text{VAR}(\sum_{j=1}^{m} X_j) = \text{COV}(\sum_{j=1}^{m} X_j, \sum_{j=1}^{m} X_j) = \sum_{j=1}^{m} \sum_{k=1}^{m} \text{COV}(X_j, X_k) = \sum_{j=1}^{m} \text{COV}(X_j, X_j) + \sum_{j=1}^{m} \sum_{k \neq j=1}^{m} \text{COV}(X_j, X_k) = \sum_{j=1}^{m} \text{VAR}(X_j) + 2\sum_{j=1}^{m} \sum_{k<j} \text{COV}(X_j, X_k)$.

Theorem 1.14 *If X_i and Y_i are pair-wise independent, and $U = c_1 * X_1 + c_2 * X_2 + \cdots + c_n * X_n$ and $V = d_1 * Y_1 + d_2 * Y_2 + \cdots + d_n * Y_n$, then $\text{COV}(U, V) = \sum_{k=1}^{n} c_k * d_k * \text{COV}(X_k, Y_k)$.*

Proof Consider $U*V = \sum_{i=1}^{n} \sum_{j=1}^{n} c_i d_j X_i Y_j$. Separate the index var into two groups as $i = j$ and $i \neq j$ and write this as $\sum_{k=1}^{n} c_k * d_k * X_k Y_k + \sum_{i=1}^{n} \sum_{j \neq i=1}^{n} c_i d_j X_i Y_j$. Take covariance of both sides to get $\text{Cov}(U, V) = \text{Cov}[\sum_{k=1}^{n} c_k * d_k * X_k Y_k] + \text{Cov}[\sum_{i=1}^{n} \sum_{j \neq i=1}^{n} c_i d_j X_i Y_j]$. As X_i and Y_i are pair-wise independent, the second sum is zero. Taking Covariance inside the summation in the first term gives the result.

1.4.6 Moments as Expected Values

The arithmetic mean of a RV is the first *raw* or *uncentered* moment, which is denoted as $\mu = \text{E}(X)$. We call $\text{E}(X^k)$ as the kth raw moment and denote it as m_k; and $\text{E}((X - \mu)^k)$ as the kth central moment μ_k. Here μ is called the pivot. Theoretically, the pivot can be any nonzero constant, so long as the expected value exists. The moment generating function (MGF) of a RV is $\text{E}(e^{tx}) = \sum_x e^{tx} f(x)$ for X discrete, and $\int_x e^{tx} f(x)dx$ for X continuous. The characteristic function (ChF) is $\text{E}(e^{itx})$ where i is the imaginary constant. The MGF may not exist for some RVs, but the ChF always exist. By expanding $(X - \mu)^k$ using binomial theorem, it is possible to express the central moments in terms of raw moments as follows:

$$\mu_k = \text{E}(X^k) = \sum_{j=0}^{k} \binom{k}{j}(-\mu)^{k-j} m_j. \tag{1.18}$$

When k is even, $(X - \mu)^k = (\mu - X)^k$ so that $\mu_k = \sum_{j=0}^{k}(-1)^{k-j}\binom{k}{j}\mu^j m_{k-j}$.

Theorem 1.15 *The change of origin and scale transformation yields* $\mu_r(c*X+b)=c^r * \mu_r(X)$.

By definition $\mu_r(c*X + b) = E[cX + b - E(cX + b)]^r$. *The* $+b$ *and* $-b$ *cancels out giving* $E[cX - E(cX)]^r = c^r * \mu_r(X)$. *Replace b by* $-b$ *to get a similar result.*

Theorem 1.16 *Prove that* $E[cX + b]^n = \mu'_r(c*X + b) = \sum_{k=0}^{n} \binom{n}{k} c^{n-k} b^k E[X^{n-k}]$ *where r is a positive integer.*

Here μ'_r *denotes the raw moments given by* $E[cX + b]^r$. *As r is a positive integer, we could expand this as a power series to get* $E[\sum_{k=0}^{r} \binom{r}{k} b^k (cX)^{r-k}]$. *Take* c^{r-k} *as a constant outside and operate the expectation on each term to get the result.*

Example MGF

If $f(x; a, b) = C\, x^{a-1} e^{-(x/b)^a}$ prove that $C = a/b^a$. Find the CDF and MGF, and obtain the first two moments.

Solution Put $y = x^a$ so that $dy = a x^{a-1}$. As $x \geq 0$, range of y remains the same. Thus $f(y; a, b) = (C/a)\int_0^{\infty} \exp(-y/b^a)$. Using gamma integral this becomes $(C/a)*b^a$ from which $C = a/b^a$. As $M_x(t) = E(e^{tx})$, we could expand $E(e^{tx})$ as an infinite series, combine each power with x^{a-1} and then put $y = x^a$ in each term to get a term-by-term integral.

$E[[X(1 - X)]^k]$ of Beta distribution

If $X \sim \text{BETA}(a, b)$ where $a < b$, find $E[[X(1 - X)]^k]$

Solution $E[[X(1 - X)]^k] = C*\int_0^1 x^k (1 - x)^k x^{a-1}(1 - x)^{b-1} dx$ where $C = 1/B(a, b)$. Combine like terms to get $C*\int_0^1 x^{a+k-1}(1 - x)^{b+k-1} dx = B(a + k, b + k)/B(a, b)$.

MGF of geometric distribution

Find the MGF and variance of the geometric distribution $f(x) = (1 - e^{-\lambda}) e^{-\lambda x}$.

Solution Here $p = e^{-\lambda}$ and $q = 1 - e^{-\lambda}$. Substitute these values for the expression for moments (given below). $\mu = q/p = 1 - e^{-\lambda}/e^{-\lambda}$, and second raw moment is $1 - e^{-\lambda}/e^{-2\lambda} + (1 - e^{-2\lambda})/e^{-2\lambda}$. From this the variance follows as $1 - e^{-\lambda}/e^{-2\lambda}$.

Mean of geometric distribution

For the Geometric distribution with PMF f(x) = $q^{x-1}p$, find E(X) and E(X^2) and deduce
Var(X). What is the MD?

Solution The mean is given by

$$\mu = \sum_{x=0}^{\infty} xq^x p = p[q + 2q^2 + 3q^3 + \cdots] = pq[1-q]^{-2} = pq/p^2 = q/p. \quad (1.19)$$

If the PMF is taken as f(x; p)=$q^{x-1}p$, the mean is 1/p. E(X^2) = $\sum_{x=0}^{\infty} x^2 q^x p$. Replace p by
1–q and split it into two series: E(X^2) = $\sum_{x=1}^{\infty}(1-q)x^2 q^x = \sum_{x=1}^{\infty} x^2 q^x - q\sum_{x=1}^{\infty} x^2 q^x$.
Expand term by term to get q–q^2 + 4q^2 – 4q^3 + 9q^3 – 9q^4 + $\cdots$. This simplifies to q +
3q^2 + 5q^3 + 7 q^4 + $\cdots$. Write this as q[1 + 2q + 3q^2 + 4q^3 + $\cdots$] + q^2[1+2q + 3q^2 +
$\cdots$] = q/p^2 + q^2/p^2. Now use σ^2 = E(X^2) – [E(X)]^2 to get q/p^2. The PDF of Y as $q^{y-1}p$
is related through a change of origin transformation Y = X – 1. This simply displaces the
distribution to the left or right. Using E(Y) = E(X)–1, we get E(Y) = (1/p–1) = (1–p)/p =
q/p. Variance remains the same because V(Y) = V(X). Now consider the mean deviation
(MD) using the power method given in [Chattamvelli and Shanmugam (2021)] we get
MD = 2$\sum_{x=ll}^{\mu-1}$ F(x) = 2$\sum_{x=0}^{c}[1-q^{x+1}]$, where c = $\lfloor q/p \rfloor$ – 1.

Absolute deviation from median

For any continuous distribution, prove that E|X – c| is minimum when c is the median.

Solution Let X be discrete. By definition, E|X – c| = $\sum_{x_i<c}(c-x_i)f(x) + \sum_{x_i>c}(x_i-c)$
f(x). Perturb the constant c by a small amount δc so that c = c – δc. The net change
is then Δ = –$\sum_{x_i<c}(\delta c)f(x) + \sum_{x_i>c}(\delta c)f(x)$. Taking the constant δc outside the summation, we get Δ = δc$\left[\sum_{x_i>c}f(x) - \sum_{x_i<c}f(x)\right]$. If c is the median, then the expression in the
square brackets is zero (because the median divides the total frequency into equal parts).
Thus the result. If X is continuous, we could write

$$|X-c| - |X-M| = \begin{cases} c-M & \text{for } x<c; \\ 2(X-c)+c-M & \text{for } c \le x \le M; \\ M \quad c & \text{for } X>M. \end{cases}$$

Whereas the mean balances the data above and below it in terms of the magnitudes of obser-
vations, the median balances the frequency (count) of data above and below it, irrespective

Table 1.4 Sum of the numbers in two dice throws

2	3	4	5	6	7	8	9	10	11	12
1/36	2/36	3/36	4/36	5/36	6/36	5/36	4/36	3/36	2/36	1/36

of their magnitudes (here we are assuming that the median for even sample size is the mean of the middle (sorted) sample values).

Sum of the numbers on two unbiased dice

Let X be a random variable that denotes the sum of the numbers that shows up when two unbiased dice are thrown. Define the PMF of X and find its expected value E(X). Does E(E(X)) exist?

Solution Obviously X can take the values between 2 and 12. The distribution is shown in the Table 1.4. From this we get E(X) = 1/36[2 + 3*2 + 4*3 + ...] = 2 + 6 + 12 + 20 + 30 + 42 + 40 + 36 + 30 + 22 + 12 = 252/36 = 7. $E(X^2)$ = 2.77778 + 5 + 8.166667 + 8.888889 + 9 + 8.33333 + 6.72222 + 4 = 54.83333, so that V(X) = $E(X^2)$–$E(X)^2$ = 54.83333–49 = 5.83333. E(E(X)) being the expectation of a constant exists and is 7 (Table 1.4).

MGF of size-biased exponential distribution

If f(x) = K x exp(–x) for x ≥ 0, find K, E(X) and the MGF.

Solution Integrate f(x) from 0 to ∞ to get $K \int_0^\infty x^{2-1} \exp(-x) dx = K\Gamma(2) = 1$. As $\Gamma(2) = 1!$ = 1, we get K = 1. E(X) = $\int_0^\infty x^{3-1} \exp(-x) dx = \Gamma(3) = 2$. Next consider $M_x(t) = E(e^{tx}) = \int_0^\infty e^{tx} x^{2-1} \exp(-x) dx$. Combine like terms to get $M_x(t) = \int_0^\infty x^{2-1} e^{-(1-t)x} dx = \Gamma(2)/(1-t)^2 = (1-t)^{-2} = 1 + 2t + 3t^2 + 4t^3 + \cdots$.

Expectation of Ratio

Find $E(\frac{1\pm X}{1+Y})$ and $E(\frac{1\pm X^2}{1+Y^2})$ for the random variable f(x, y) = x + y, 0 < x, y < 1.

Solution We are given that f(x, y) = x + y for 0 < x < 1, 0 < y < 1. From this f(x) = $\int_{y=0}^{1}(x+y)dy$ = x + 1/2. Due to symmetry, f(y) = y + 1/2. As f(x, y)≠f(x)*f(y), X and Y are dependent. Since the numerator and denominator are dependent, we use the following formula

$$E[X/Y] \simeq \mu_x/\mu_y \left[1 + \text{Var(Y)}/\mu_y^2 - \text{Cov(X, Y)}/(\mu_x\mu_y)\right], \qquad (1.20)$$

where $\mu_x = \int_0^1 x(x + 1/2)dx = 1/3 + 1/4 = 7/12$. Due to symmetry $\mu_y = 7/12$. $E(Y^2) = \int_0^1 y^2(y + 1/2)dy = 1/4 + 1/6 = 10/24 = 5/12$, so that $V(Y) = 5/12 - 7/12*7/12 = 11/144$. $E(XY) = \int_0^1 \int_0^1 xy(x + y)dxdy = 1/6 + 1/6 = 1/3$. As the numerator is $(1 \pm X)$, we have to replace μ_x by $(1 \pm E(X)) = (1 \pm 7/12, \mu_y = E(1 + Y) = 1 + 7/12 = 19/12$, $\text{Var}(1 + Y) = \text{Var(Y)} = 11/144$, $\text{Cov}(1 \pm X, 1 + Y) = \text{Cov(XY)} = E(XY)-E(X)*E(Y) = 1/3-7/12*7/12 = -1/144$. Substitute these values to get the result. For part 2, proceed exactly as above.

Prove $E(h(x)) \geq h(E[x])$

If the second derivative of h(x) is positive, prove $E[h(x)] \geq h(E[x])$.

Solution Denote E(X) by μ, a constant, and note that $E(X-\mu) = 0$. Expand h(x) as a Taylor series to obtain

$$h(x) = h(\mu) + (x - \mu)\left[\frac{\partial}{\partial x}h(x)\right]_{x=\mu} + (x - \mu)^2/2!\left[\frac{\partial^2}{\partial x^2}h(x)\right]_{x=\mu} + \cdots \qquad (1.21)$$

Take expectation of both sides, and use $E(X-\mu) = 0$ on the RHS to get $E(h(x)) = h(\mu) + 0 + E(x - \mu)^2/2\left[\frac{\partial^2}{\partial x^2}h(x)\right]_{x=\mu} + \cdots$ As $E(x - \mu)^2/2 \geq 0$, and using the given condition that the second derivative of h(x) is positive, it follows that $E(h(x)) \geq h(\mu) = h(E(x))$.

Pr[X < Y]

If X and Y are IID EXP(λ_i) with respective PDF f(x) and g(y) for i = 1, 2 prove that

$$\text{Pr(X < Y)} = \int_0^\infty f(x)[1 - G(x)]dx = 1 - \int_0^\infty g(y)[1 - F(y)]dy = \lambda_1/(\lambda_1 + \lambda_2)$$

Solution The PDF of X is $f(x) = \lambda_1 \exp(-\lambda_1 x)$ and Y is $g(y) = \lambda_2 \exp(-\lambda_2 y)$. The CDF of Y is $G(y) = 1 - \exp(-\lambda_2 y)$, so that $1 - G(y) - \exp(-\lambda_2 y)$. Hence $\int_0^\infty f(x)[1 - G(x)]dx = \int_0^\infty \lambda_1 \exp(-\lambda_1 x) \exp(-\lambda_2 x)dx = \lambda_1/(\lambda_1 + \lambda_2)$. Similarly $1 - \int_0^\infty g(y)[1 - F(y)]dy = \lambda_1/(\lambda_1 + \lambda_2)$.

Expected number of devices in working condition

The lifetime of a device in years is distributed as EXP(λ) where $\lambda = 1/8$. If n such devices are put together in a satellite, find the following:– (i) probability that half or more of the devices are in good working condition after 5 years. (ii) Expected number of devices in working condition after 8 years.

Solution Put t=5 to get the probability that any device is working after 5 years as $\lambda \exp(-5 * \lambda) = 0.0669$. Probability that it is not working is $1 - \lambda \exp(-5\lambda) = 0.9331$. As there are n such devices, probability that at least half of the devices are working is $\sum_{k=n/2}^{n} \binom{n}{k}[0.0669.]^{k}[0.9331]^{n-k}$. Using the relationship between binomial SF and incomplete beta function, this can be written as $I_c(n/2, n/2 + 1)$ where $c = 0.0669$. For case (ii), we need to find the expected value after 8 years. The probability of good working condition is $\lambda \exp(-8\lambda)$. Number of devices in working condition is a binomial variate, so that the expected value is $np = n\lambda * \exp(-8\lambda) = 0.04598493*n$.

1.5   Conditional Expectations

Conditional expectation is a useful concept that defines the expected value of a RV or function thereof by conditioning one or more dependent variables. Measures of risk frequency (stock markets, finance and insurance) uses conditional tail expectations, $E(X|Y>y)$ and $E(Y|X>x)$. Conditional expectation can also be defined in terms of conditional density functions. The conditional expectation considers a non-null subset of RVs by fixing some other RVs as constant.

$$E[Y|X = x] = \begin{cases} \sum_{x=-\infty}^{\infty} y f_y(Y|X = x) & \text{if } Y \text{ is discrete;} \\ \int_{y=-\infty}^{\infty} y f_y(Y|X = x) dy & \text{if } Y \text{ is continuous.} \end{cases}$$

Thus the conditional expected value of Y for a given value of X = x is the mean of Y computed relative to the conditional distribution, which is a function of x.

Theorem 1.17 *Show that E(Y) = E[E(Y|X)] if X and Y are independent.*

Proof : For simplicity assume that X and Y are continuous. Consider the RHS. $E[E(Y|X)] = \int_{x=-\infty}^{\infty} E[Y|X = x] f_X(x) \, dx$. Here we have expanded the outer expectation operator. Next expand the inner expectation operator to get

$$E[E(Y|X)] = \int_{x=-\infty}^{\infty} \left(\int_{y=-\infty}^{\infty} y f_{X,Y}(x, y)/f_X(x) dy \right) f_X(x) dx.$$

As $f_X(x)$ inside the inner integral is a constant while integrating wrt y, this cancels out from the numerator and denominator to get

$$\int_{x=-\infty}^{\infty} \int_{y=-\infty}^{\infty} y\, f_{X,Y}(x, y)dxdy.$$

As X and Y are independent, the density function $f_{X,Y}(x, y)$ factors into $f_X(x) * f_Y(y)$. Integrate out f(x) over its entire range to unity, and the remaining expression becomes $\int_{y=-\infty}^{\infty} f_Y(y)dy = E[Y]$, which is the LHS.

Theorem 1.18 *Show that E(XY) = E[E(XY|X)].*

Proof As before, assume that X and Y are continuous. Consider the RHS.

$$E[XY|X=x] = \int_{y=-\infty}^{\infty} xy f_{X,Y}(x, y)/f_X(x)dy \qquad (1.22)$$

As x is a constant inside the integral over y, this becomes

$$x \int_{y=-\infty}^{\infty} y f_{X,Y}(x, y)/f_X(x)dy.$$

This integral was shown above as $E[Y|X = x]$. Expand the outer expectation operator on the RHS as $E[E(XY|X)] = \int_{x=-\infty}^{\infty} E[XY|X = x] f_X(x)dx$. Next expand the inner integral also to get

$$E[E(XY|X)] = \int_{x=-\infty}^{\infty} \left(\int_{y=-\infty}^{\infty} xy\, f_{X,Y}(x, y)/f_X(x)dy \right) f_X(x)dx.$$

As $f_X(x)$ inside the inner integral is a constant while integrating wrt y, this cancels out from the numerator and denominator to get

$$E[E(XY|X)] = \int_{x=-\infty}^{\infty} \int_{y=-\infty}^{\infty} xy\, f_{X,Y}(x, y)dxdy = E(XY).$$

Theorem 1.19 *Show that E(Y + Z | X) = E(Y | X) + E(Z | X).*

Proof We will prove the result for the continuous case. By the above definition, $E(Y + Z | X) = \int_{y=-\infty}^{\infty} (y + z) f_y(Y|X = x)dy = \int_{y=-\infty}^{\infty} y f_y(Y|X = x)dy + \int_{y=-\infty}^{\infty} z f_y(Y|X = x)dy = E(Y | X) + E(Z | X)$.

Random sum of IID random variables

Suppose $X_1, X_2, \ldots, X_n$ are IID RVs, with $E(X_i) = \mu$ and $var(X_i) = \sigma^2$. Define Y = $X_1 + X_2 + \cdots + X_N$ where N is another RV independent of X. Prove that $E(Y|N) = N\mu$, and $\text{Var}(Y|N) = N\sigma^2$. Use $E(Y) = E[E(Y|N)]$ to show that $E(Y) = \nu\mu$ where $\nu = E(N)$. Show that the unconditional variance of Y is $\nu\sigma^2 + \mu^2\delta^2$ where δ^2 is variance of N.

Solution As $Y = X_1 + X_2 + \cdots + X_N$, we get $E(Y) = E[E(Y|N)] = E[N\mu] = \mu\,E[N] = \mu\nu$. Next use $\text{Var}(Y) = \text{Var}[E(Y|N)] + E[\text{Var}(Y|N)]$. $\text{Var}(Y|N) = N\sigma^2$, so that $E[\text{Var}(Y|N)] = E[N\sigma^2] = \sigma^2 E[N]$. Similarly $E[Y|N] = N\mu$ so that $\text{Var}(E[Y|N]) = \text{Var}(N\mu) = \mu^2\text{Var}(N)$. Substitute the values to get the result.

Mean of noncentral beta distribution

Find the mean of noncentral beta distribution using conditional expectation.

Solution The noncentral beta distribution is an infinite sum of Poisson weighted central beta distributions [Chattamvelli and Shanmugam (1998)]. Depending on whether the central beta distribution is of first or second kind, there exist noncentral beta distribution (NCB) of two kinds [Chattamvelli (1995)]. Symbolically, NCB of first kind has CDF $I_x(a, b; \lambda) \equiv I_x(a + N, b)$, where $N \sim P(\frac{\lambda}{2})$ has a Poisson distribution. Hence conditional on N, the RV X has a central beta distribution of first kind. From this, an expression for the mean is easily obtained as follows.

$$E(X) = E\left[E(X|N)\right] = E\left[(a + N)/(a + b + N)\right], \tag{1.23}$$

where we have used the fact that the mean of a beta distribution of first kind is $a/(a+b)$. Write the numerator $(a + N)$ as $(a + b + N) - b$ and simplify to get the RHS as $1 - b*E[1/(a + b + N)]$. As N is Poisson distributed, the expression in the square bracket is the first inverse moment of a displaced Poisson distribution, which is given as $E\left(\frac{1}{A+N}\right) = \frac{e^{-\lambda/2}}{A} {}_1F_1[A, A + 1; \lambda/2]$ where $A = a + b$, and ${}_1F_1[A, A + 1; \lambda/2]$ is the confluent hypergeometric function. This gives an exact expression as

$$\mu = 1 - (b/A)\exp(-\lambda/2) \ {}_1F_1[A, A + 1; \lambda/2]), \quad \text{where } A = a + b. \tag{1.24}$$

Since the numerator and the denominator of (1.23) are dependent, we use the following formula

$$E\left[X/Y\right] \simeq \mu_x/\mu_y \left[1 + \text{Var}(Y)/\mu_y^2 - \text{Cov}(X,Y)/(\mu_x\mu_y)\right],$$

where μ_x and μ_y denote the mean of X and Y respectively. Here $\mu_x = E[b] = b$, $\mu_y = E(a + b + N) = (a + b + \lambda/2)$, $\text{Cov}(X, Y) = 0$, $\text{Var}(X) = 0$, $\text{Var}(Y) = \text{Var}(a + b + N) = \text{Var}(N)$ (using Sect. 5) $= \lambda/2$. Hence (1.23) becomes

$$\mu = E(X) \simeq 1 - (b/C)[1 + D/C^2], \tag{1.25}$$

Table 1.5 Mean of noncentral beta distribution

a	b	λ	D=$\lambda/2$	C = a + b + D	$1 - \frac{b}{C}*(1 + \frac{D}{C^2})$	Actual mean	Difference
1	2	0.5	0.25	3.25	0.37005007	0.37300047	0.00295
3	3	2	1.0	7	0.56268220	0.56340118	0.000719
10	2	0.5	0.25	12.25	0.83646269	0.83648217	0.00002
2	10	4	2.0	14	0.27842566	0.27872571	0.00030
10	10	6	3.0	23	0.56275171	0.56207801	−0.00067
5	5	20	10.0	20	0.74375000	0.74359903	−0.00015
12	10	30	15.0	37	0.72676841	0.72675244	−0.00002
20	10	40	20.0	50	0.79840000	0.79839399	−0.00001

where D = $\lambda/2$, and C = a + b + D [Chattamvelli and Shanmugam (1998)]. Some approximations are given in Table 1.5, where the actual mean is an infinite sum as

$$\mu = \sum_{k=0}^{\infty} e^{-\lambda/2}(\lambda/2)^k/k![(a+k)/(a+b+k)]. \tag{1.26}$$

The difference between actual and approximate values are given in the last column. The biggest advantage of (1.25) is that it takes only 9 arithmetic operations (including the computation of D=$\lambda/2$ and C once) whereas (1.26) takes a large number of operations when λ is large.

Mean of noncentral χ^2 distribution

Find the mean of noncentral chi-square distribution using conditional expectation.

Solution Let Y be distributed as noncentral chi-square. As this is a Poisson weighted central chi-square distribution we write $Y \sim \chi^2_{n+2N}$ where N has Poisson distribution with parameter $\lambda/2$. From this we get E(N) = $\lambda/2$.

$$E(Y) = E[E(Y|N)] = E(n + 2N) = n + \lambda. \tag{1.27}$$

Mean of noncentral F distribution

Find the mean of noncentral F distribution.

Solution The noncentral F distribution is the distribution of the scaled ratio of a non-central $\chi^2(\lambda)$ over an independent central χ^2 distribution. Symbolically $F(p,q,\lambda) = (q/p)\,\chi^2_{p+2N}/\chi^2_q$ where p and q are the DoF of numerator and denominator χ^2, and λ is the noncentrality parameter. This may be written as $Z = F(p,q,\lambda) \sim \frac{(p+2N)\chi^2_{p+2N}/(p+2N)}{p\chi^2_q/q}$. Conditional on N, the noncentral F distribution is a multiple of central F distribution ([Chattamvelli and Jones (1995); Raja Rao (1986)]). We write this as $F(p,q,\lambda) \sim \frac{p+2N}{p} F_{p+2N,q}$. The moments follow by the same argument as

$$E(Z) = E[E(Z|N)] = E\left[\frac{p+2N}{p}\frac{q}{q-2}\right]$$
$$= \frac{q}{q-2}\frac{p+\lambda}{p}, q > 2$$

where we have used the fact that the mean of central F(p, q) is q/(q–2).

1.5.1 Conditional Variances

The variance of a RV, conditionally on another variable or on the count of IID RVs occur in several applications. The conditional variance of Y for a given X is var(Y | X) = E{[Y – E(Y | X)]²| X}. Expanding the quadratic and taking term by term expectation results in Var(Y | X) = E(Y^2| X) – [E(Y | X)]².

Theorem 1.20 *Let N be an integer valued RV that takes values ≥ 1. Let $X_1, X_2, \ldots X_N$ be N independent identically distributed (IID) RVs. Define $S_N = X_1 + X_2 + \cdots + X_N$. Then (i) E(S) = E(X) E(N), provided the expectations exist, (ii) $P_{S_N}(t) = P_N(P_X(t))$, and $M_{S_N}(t) = P_N(K_X(t)) = (K_X(t))^N$ due to independence.*

Proof Assume that N is fixed. Then E(S) = E($X_1 + X_2 + \cdots + X_N$) = E(X_1) + E(X_2) + $\cdots$ + E(X_N). As each of the X_i's are IID, the above becomes E(S) = N E(X). Now allow N to vary in its range and take the expectation of both sides to get E(S) = E(N)E(X) (because E(E(S)) = E(S) and E(E(X)) = E(X)).

To prove $P_{S_N}(t) = P_N(P_X(t))$, we proceed as above and assume that N is held constant. Then $P_{S_N}(t) = E(t^{S_N}) = E(t^{X_1+X_2+\cdots+X_N}) = E(t^{X_1})E(t^{X_2})\cdots E(t^{X_N}) = E(t^X)^N = [P_X(t)]^N$. Now consider N as a discrete variate and $t' = P_X(t)$ as a constant. Take expectation of both sides to get the desired result. Next consider $M_{S_N}(t) = E(e^{tS_N}) = E(e^{t[X_1+X_2+\cdots+X_N]}) = M_{X_1}(t) * M_{X_2}(t) * \cdots * M_{X_N}(t) = [M_X(t)]^N$. Taking log of both sides, we get $K_{S_N}(t) = \log(M_{S_N}(t)) = N * \log(M_X(t)) = NK_X(t)$. Now allow N to vary to get the result.

1.5.2   Law of Conditional Variances

Theorem 1.21 *The unconditional variance can be expressed in terms of conditional variances as $V(X) = E[V(X|Y)] + V[E[X|Y]]$ where $V(X) = Variance(X)$, assuming that the variances exist.*

Proof Subtract and add $E[X|Y]$, and write $X–E[X] = (X–E[X|Y]) + (E[X|Y]–E[X])$. Square both sides and take expectation of each term to get

$$E[X–E[X]]^2 = V(X) = E(X - E[X \mid Y])^2 + E(E[X \mid Y] - E[X])^2 +$$
$$2E(X - E[X \mid Y])(E[X \mid Y] - E[X]) = (1) + (2) + (3) \text{ (say)}. \qquad (1.28)$$

As $E(E[X|Y]) = E(X)$, the last term (3) is zero. Substitute $E(X) = E(E[X|Y])$ in the second term $E(E[X \mid Y] - E[X])^2$ to get $(2) = E(E[X \mid Y] - E(E[X|Y]))^2$. As this is the expectation of the squared deviation of $E[X \mid Y]$ from its mean, it is $Var(E[X|Y])$. Symbolically, $(2) = Var(E[X|Y])$.

Using the law of total expectation we have $V(X|Y) = E[X^2|Y] - E[X|Y]^2$. Take expectation of both sides to get $E[V(X|Y)] = E\{E[X^2|Y] - E[E[X|Y]^2]\}$. Write the first term $E(X - E[X \mid Y])^2$ in (1.28) as $E\{E(X - E[X \mid Y])^2|Y\}$. Expand the quadratic and take term by term expectation to get $E\{E[X^2|Y]\} - 2E\{E(X)E[X|Y]\} + E\{E[X|Y]^2\}$. Substitute $E(X) = E(E[X|Y])$ in the second term, and cancel out the third term. This reduces to $E\{E[X^2|Y]\} - E\{E[X|Y]^2\}$ showing that it is the expected value of $V[X|Y]$. Symbolically, $(1) = E[V(X|Y)]$. Substitute for (1) and (2) in (1.28) to get the result (see Table 1.6).

If X and Y are continuous RVs and $g : \mathbb{R} \to \mathbb{R}$ is a continuous bounded integrable function, then $E[g(x)|y] = \int_x g(x) f_{X|y}(x|y)dx$ is called the conditional expectation of $g(x)$ given y. Replace integration by summation to get an analogous result for discrete RVs.

Variance of noncentral chi-square

Find the variance of noncentral chi-square distribution using conditional expectation.

Solution We know that the noncentral chi-square distribution (Y) is a Poisson weighted linear combination of independent central chi-square distributions (X). This allows us to write it as $Y \sim X_{p+2N}$ where conditional on N, X is a central chi-square distribution. For convenience let the DoF of central chi-square be denoted by p. Then

$$V(Y) = V[E(Y|N)] + E[V(Y|N)] = V(p + 2N) + E(2p + 4N)$$
$$= 4V(N) + 2p + 4E(N) = 2p + 4\lambda \qquad (1.29)$$

Table 1.6 Summary table of expressions for variance

Using	Expression	Comment
Definition	$EX^2 - E(X)^2$	$E(X)^2$, $E(X)$ finite
Var(Y | X)	$E(Y^2\|X) - [E(Y\|X)]^2$	Conditional on X
V(X)	$E[V(X\|Y)] + V[E(X\|Y)]$	Unconditional & Conditional
E[X(X–1)]	$E[X(X-1)] + E(X)[1-E(X)]$	X discr., E[X(X–1)] finite
CDF	$\int_0^\infty 2x[1 - F_X(x) + F_X(-x)]dx - \mu_X^2$	
$P_x(t)$	$P_x''(1) + P_x'(1) - [P_x'(1)]^2$	PGF
$M_x(t)$	$M_x''(0) - [M_x'(0)]^2$	MGF
$K_x(t)$	$K_x''(0)$	$K_x'(0) = \mu$
$\ln(F_X(x))$	$[\ln(F_X(x))]'\|_{x=1} + [\ln(F_X(x))]''\|_{x=1}$	F is CDF
$\phi_x(t)$	$\phi_x''(0) - [\phi_x'(0)]^2$	ChF
$FM_x(t)$	$FM_x''(0) + FM_x'(0) - [FM_x'(0)]^2$	FMGF
Tails‡	$[2/(cf_m)]\int_{ll}^{\mu} F(x)dx$ (Continuous)	$[2/(cf_m)]\sum_{ll}^{\lfloor\mu\rfloor} F(x)$ (Discrete)

see [Wilf (1994)]. ‡See power method in [Chattamvelli and Shanmugam (2021)]

where we have used the facts that $V(c + b*X) = b^2 V(X)$, and $E(X) = V(X) = \lambda$ for a Poisson distribution ([Raja Rao (1986); Chattamvelli and Shanmugam (1998)]).

1.6   Inverse Moments

The definition of ordinary moments can be extended to the case where the order is a negative integer as follows:

$$E(1/x^k) = \begin{cases} \sum_{j=-\infty}^{\infty} x_j^{-k} p_j & \text{if } X \text{ is discrete}; \\ \int_{-\infty}^{\infty} x^{-k} f(x)dx & \text{if } X \text{ is continuous}. \end{cases}$$

A necessary condition for the existence of the first inverse moment is that f(0) = 0. For instance, the Poisson distribution has $p(x = 0) = e^{-\lambda}\lambda^0/0! = e^{-\lambda}$, which is nonzero $\forall\lambda$. Hence the first inverse moment does not exist. But there are a large number of distributions that satisfy the necessary condition. Examples are chi-square (and gamma), Snedecor's F, beta and Weibull distributions. The exponent k is integer in most of the applications of inverse moments. But inverse moments could also be defined for fractional k (called fractional inverse moments). See [Chattamvelli and Jones (1995)] for recurrences on inverse moments of noncentral distributions, from which recurrences for respective central moments ($\lambda = 0$) follow easily.

Problem 1.3 Show that for positive random variables $E(1/X) \geq E(X^{a-1})/E(X^a)$ for a>0. Hint: Show that $E[(X-\mu)(1/X-1/\mu)] \leq 0$ where $\mu = E(X)$.

Inverse moment of central χ^2 distribution

Find first inverse moment of central χ^2 distribution.

Solution By definition, $E(1/X) =$

$$K \int_0^\infty (1/x)x^{n/2-1}e^{-x/2}dx = K \int_0^\infty x^{n/2-2}e^{-x/2}dx = K\Gamma(n/2-1)2^{n/2-1} \quad (1.30)$$

where $K = 1/(2^{n/2}\Gamma(n/2))$. This simplifies to $1/(n-2)$.

1.7   Incomplete Moments

Ordinary and central moments discussed above are defined for the entire range of the RV X. There are several applications when the summation or integration is carried out partially over the range of X. The omitted range can either be in the left tail or in the right tail. We define the first incomplete moment as $E_I(X) = \sum_{x=k}^\infty xf(x)$ for X discrete and $E(X) = \int_k^{ul} xf(x)dx$ for X continuous.

1.8   Distances as Expected Values

Statistical distances can be expressed as expected values. Consider two real-valued RVs X and Y. The k-norm distance between them is $D_k(X, Y) = ||X - Y||^k = [E(|Y - X|^k)]^{1/k}$ (Table 1.7). This is also called the k-metric. It satisfies the following properties: (a) $D_k(X, Y) \geq 0$, (b) $D_k(X, Y) = 0$ iff X = Y, c) $D_k(X, Y) + D_k(Y, Z) \geq D_k(X, Z)$ (triangle inequality). Particular values of k give various distances like Euclidean metric, Manhattan metric etc. [Chattamvelli (2016)]. The sample analogs of these distances are used in cluster analysis as dissimilarity metrics. The above definition can be extended from scalar RVs to vectors and matrices. For instance, if X is an m × n matrix of real-valued RVs, where X_{ij} denotes the (i, j)th entry, we define E(X) as that matrix whose (i, j)th entry is $E[X_{ij}]$, provided the individual expectations exist. Using matrix commutativity, associativity etc. wrt addition, we could obtain the following results:

Table 1.7 Summary of mathematical expectation

Function	Name	Conditions
$E[X^k]$	kth raw moment	k is real, finite
$E[X^{-k}]$	kth inverse moment	k real, f(x = 0) = 0
$E[(X - \mu)^k]$	kth central moment	k is real, finite
$E|X - \mu|$	Mean deviation (about mean)	μ finite
$E[X(X-1)\cdots(X-k+1)]$	kth falling factorial moment	$k \geq 1$ is real, finite
$E[X(X+1)\cdots(X+k-1)]$	kth raising factorial moment	$k \geq 1$ is real, finite
$EX^2 - E(X)^2$	Variance	$E(X)^2, E(X)$ finite
$E[X(X-1)] + E(X)[1-E(X)]$	Variance	$EX(X-1)$ finite
$E[(X - \mu_x)(Y - \mu_y)]$	Covariance	μ_x, μ_y finite
$E(c_1 g_1(x) + c_2 g_2(x) + \cdots)$	Linear combination	c_i's $\neq 0$
$E(e^{tx})$	Moment gen. function	$-\epsilon < t < \epsilon, \epsilon > 0$
$E(e^{itx})$	Characteristic function	$-\epsilon < t < \epsilon, \epsilon > 0$
$E(t^x)$	Probability gen. function	$-\epsilon < t < \epsilon, \epsilon > 0$
$E(t^x)/(1-t)$	CDF gen. function	$0 < t < 1$
$2E(t^x)/(1-t)^2$	MD gen. function	Coeff. of $t^{\lfloor \mu \rfloor}$, Discr.
$E[(1+t)^x]$	Falling factorial MGF	$-\epsilon < t < \epsilon, \epsilon > 0$
$E[(1-t)^{-x}]$	Raising factorial MGF	$-\epsilon < t < \epsilon, \epsilon > 0$
$E(|X - Y|^k)^{1/k}$	Distance (X, Y)	k-norm
$\frac{\partial}{\partial \theta} E[h(x,\theta)] = E[\frac{\partial}{\partial \theta} h(x, \theta)]$	Derivative	Continuous in θ
$E(XY)^2 \leq E(X^2)E(Y^2)$	Cauchy-Schwartz inequality	
$E(g(x)) \geq g(E(x))$	Jensen's inequality	g(x) is convex
$P[g(X) \geq c] \leq E[g(X)]/c$	Chebychev Inequality	Random variable X

(i) E(X + Y) = E(X) + E(Y) if X and Y are compatible matrices, (ii) E(AX) = AE(X) if A is a scalar m × n matrix and X has as many rows as columns of A matrix (i.e.: A is n × p), (iii) if X and Y are independent, then E(XY) = E(X)E(Y).

1.9   Chebyshev Inequality

If X is a nonnegative random variable, then for any real number within its range $\Pr[X \geq a] \leq E(X)/a$ when E(X) is finite (Markov's inequality). Putting a=k E(X) tells us that the probability that X is k times as large as its mean is at most 1/k. This result connects the expected value of a function of a RV and the tail area of it. Let X be a RV and g(x) be a nonnegative function of it. Then the right tail area of g(X) is related to its expected value as $P[g(X) \geq c] \leq E[g(X)]/c$ (Table 1.7). This has a symmetric version that gives an upper bound for a RV or a function of it assuming a value around a symmetric interval about its mean as $\Pr(|X - \mu| \geq t\sigma) \leq 1 - 1/t^2$. This has an alternate version as

$$\Pr(|X - \mu| \geq t) \leq 1 - \sigma^2/t^2. \tag{1.31}$$

This is easy to interpret for symmetric laws but the result is valid for all distributions. However, it is a loose bound for the probability to deviate from it's mean. A much sharper bound is the Chernoff bound given by

$$\Pr(|X - \mu| \geq t\sigma) \leq \exp(-tk\sigma) E[\exp(t|x - \mu|)]. \tag{1.32}$$

For the two-sided Gaussian tail bound (1.31) simplifies to $\Pr(|X - \mu| \geq t) \leq 2 \exp(-t^2/(2\sigma^2))$. If the kth moment is finite, this becomes $\Pr(|X - \mu| \geq t) \leq E|X - \mu|^k/t^k$. A direct extension of it is the weak law of large numbers which states that $\Pr(|S_n - \mu| \geq c) \to 0$ for all $c > 0$ where $S_n = (X_1 + X_2 + \cdots + X_n)/n$ and X_k's are all IID RVs. We know that $E(S_n) = \mu$ and $\mathrm{Var}(S_n) = \sigma^2/n$. Apply Chebyshev's inequality using the alternate version to get $\Pr(|S_n - \mu| \leq c) \geq \sigma^2/(nc^2)$. As $n \to \infty$, the RHS tends to zero.

1.10   Applications

Mathematical expectation is used to predict event occurrences in the future. This often uses past data collected over a suitable interval. Examples are predicting stock prices, foreign exchange rates, insurance claims, fraudulent money transactions, etc. which are all ratio scale variables in NOIR typology (so that they can be modeled using a continuous RV). The waist-to-height ratio (WHR) used in medical sciences is better than BMI to predict susceptibility to certain diseases ([Dezfouli et al. (2023)]). Although zero is not well-defined for both of them (none can have a BMI or WHR of zero), they can be modeled using a continuous distribution that depends on the subjects of interest (distribution of them differs in various countries, genders, ethnic and age groups)[3] RVs are used in modeling probabilistic evolution of genomic sequences over time and in differential gene expression analysis.

Civil engineers use mathematical expectation and RVs to build dams and drainage systems to mitigate damages caused by natural phenomena. For example, existing dams in a locality may have a preset design capacity (maximum load such as water level it can withstand). Damages caused when water level exceeds the design capacity may depend on various design parameters. The capital investment on each design may vary. If past data on damages are available, the probabilities that waterflow will exceed the design capacity can be estimated in advance and the expected damage due to future events found.

Mechanical engineers use mathematical expectation to identify the force required to break a certain tensile specimen. The expected force required to break it may vary from

---

[3] BMI values less than 15 indicates an unusual health condition in a patient. The lowest recorded BMI is 6.7 (see https://pubmed.ncbi.nlm.nih.gov/35569150/).

one specimen to another. Thus it is found by applying various forces on the specimen and checking whether it is broken or not. If this experiment is repeated a number of times, we could assign a probability to each force applied to identify the peak force for which it will most likely break. All forces above a threshold are likely to break the specimen with certainty, so that an upper truncated distribution can be used to model it. It is also used to model the inherent variability in mechanical systems (e.g. performance degradation of components and systems due to thermal variations, wear and tear, corrosion, etc.), vibrational behavior of rotating machinery, and design controllers that can handle variations in system dynamics.

Mathematical expectation plays an important role in digital signal processing (DSP), actuarial sciences, astronomy and many other fields. For example, the average energy $\omega(t)$ of a periodic or random signal in the time domain is represented for continuous signals as $\omega(t) = \int_{-\infty}^{\infty} f(t)dt$, from which the average power of the signal over a time period t_1 to t_2 is given by

$$E[P] = \frac{1}{t_2 - t_1} \int_{t_1}^{t_2} f_1(t) f_2(t) dt = \frac{1}{T} \int_0^T f^2(t) dt, \tag{1.33}$$

if $f_1(t) = f_2(t)$ where f(t) represents the signal value as a time-varying function. As the spectra of periodic signals are more revealing in the frequency domain, most DSP applications use one of the frequency transforms like Fourier transform, cosine transform, wavelet transform etc. under the assumption that $\int_{-\infty}^{\infty} |x_T(t)|dt < \infty$ where $|x_T(t)|$ underscores that it is an RV in the time-domain. The average power in the frequency domain can then be represented by expected value as $E[|X_T(f)|] = \frac{1}{2T} \int_{-\infty}^{\infty} |X_t(f)|^2 df$. As T$\rightarrow \infty$, $E[|X_T(f)|]$ becomes stable for stationary processes and signals, resulting in power spectral density of the signal. The expectation of service time in packet-based communication engineering is known as average access delay. Similarly, the difference between delays (called delay jitter) experienced by successive data packets at a receiving station in long-distance communication is modeled using difference of random variables (usually right-skewed distributions) assuming that waiting delay is absent (packets do not queue to be serviced). It is used in actuarial sciences to model variables (such as losses, claim amounts) exceeding a set threshold in a fixed time period. Reliability analysis uses RVs to assess the probability of failure or malfunction of devices, instruments or components in continuous use over time. See [Whittle (2000)] for further examples. The characteristic function of the product of two RVs in which one of them is the standard normal and the other is one among nearly fifty distributions can be found in [Jiang and Nadarajah (2019)]. As shown in Sect. 1.2.5, $F^{-1}(x)$ is the CDF of a well-defined RV when X is defined on the unit interval. In [Jones and Strawderman (2002)], this result is extended to RVs defined on any finite interval (a, b) with CDF F(x) that F(x)–F(g(x)) as also F(g^{-1}(x))–F(x) are identically distributed where g(x) is an arbitrary continuous strictly increasing and invertible function. Conditional expectation is used to find low-order moments of mixture distributions. See [Giri (2004)] for integral representations of moments of noncentral beta distributions.

1.11 Summary

This chapter introduced the basic ideas and rules of both the mathematical expectation and conditional expectation. The population variance, covariance and moments are expressed as expected values. Arithmetic on expected values allows us to compute the mathematical expectation of functions of random variables. A large number of solved problems makes it easier for the reader to appreciate the usefulness of expectation in a variety of fields.

1.12 Exercises

Problem 1.4 Mark as True or False
(a) Expected value of a random variable always exists
(b) Expected value is unchanged by a change of scale transformation
(c) Chebychev inequality can provide an upper bound on expected values
(d) Variance of a distribution defined in [0, 1] can be >1
(e) $E(1/X) \geq 1/E(X)$ for all random variables X
(f) $|E[X]| \leq \sqrt{E[X^2]}$ for all random variables X
(g) If $X \leq Y$, then $E[X] \leq E[Y]$
(h) Expectation "E" is a linear and monotone operator.

Problem 1.5 A———assigns a value to each element of the sample space (a) generating function (b) random variable (c) cumulant (d) expected value.

Problem 1.6 You have an unbiased coin and an unbiased die with faces marked 1 to 6. You first throws the die and observe the number that shows up at the top. If it is an even number, you throw the die a second time. If first throw is an odd number, you throw the coin once. Describe the sample space for this random experiment.

Problem 1.7 You have a biased coin with probability of Head $p < 1/2$, and an unbiased die with faces marked 1–6. The coin is tossed (say x times) until a Head shows up. Then the die is thrown x times. Let Y denote the sum of the number that appears at the top. Find the joint distribution of X and Y and PMF of Y.

Problem 1.8 An unbiased die has the property that an odd number is twice more likely to occur on the top than an even number. Let an event V denote the number shown on the top when it is tossed. Find E(V).

Problem 1.9 An unbiased die has the property that an odd number is twice more likely to occur on the top than an even number. Define an event T as the number shown on the top when it is tossed is less than 4. Find E(T).

Problem 1.10 If $X_1, X_2, \ldots, X_n$ are IID, and $Y = \sum_{i=1}^{n} X_i$ prove that $M_Y(t) = \prod_{i=1}^{n} M_{X_i}(t)$.

Problem 1.11 If $p(x=0)=q^2$, $p(x=1)=2pq$, $p(x=2)=p^2$, find the CDF and PGF of X.

Problem 1.12 Prove that $Pr[X>a, Y>b] = 1 - F_X(a) - F_Y(b) + F_{X,Y}(a, b)$ where F() denotes the CDF, and $F_{X,Y}(a, b)$ is the joint CDF.

Problem 1.13 Prove that $Pr[x_1 < X \le x_2, y_1 < Y \le y_2] = F(x_2, y_2) - F(x_1, y_2) - F(x_2, y_1) + F(x_1, y_1)$.

Problem 1.14 Prove that the characteristic function of U(−1, +1) is sin(t)/t.

Problem 1.15 The current flow I through a resistor fluctuates according to the arcsin law $f(x) = \dfrac{1}{\pi \sqrt{(1-x^2)}}$ for $-1 < x < 1$. Find the expected value of the Power = $R * I^2$ where R is the resistance (given).

Problem 1.16 Prove that $c = E(X)$ minimizes the expression $E(x-c)^2$.

Problem 1.17 Prove that $c = \text{Median}(X)$ minimizes the expression $E|x - c|$.

Problem 1.18 Show that all cumulants except the first one vanish for a symmetric distribution.

Problem 1.19 If X is a negative random variable (values of x are always <0), prove that $E(X) = \int_0^\infty -F(-x)dx$.

Problem 1.20 If X is a non-negative discrete random variable, prove that $E(X^2) = \sum_{k=1}^{\infty} (2k+1) P(X > k)$.

Problem 1.21 If $X > 0$, prove that $E[1/X] \ge 1/E[X]$, if each expectation exists.

Problem 1.22 Prove that $COV(X, Y-Z) = COV(X, Y) - COV(X, Z)$.

Problem 1.23 If X and Y are two IID random variables, prove that $E[\max(X, Y)] + E[\min(X, Y)] = E[X] + E[Y]$.

Problem 1.24 Prove that $E(X^2) = \int_0^\infty Pr(X > t) F(t) dt = 2 \int_0^\infty t S_x(t) dt$ where $S_x(t)$ is the survival function.

Problem 1.25 Consider a game in which a fair die (marked 1–6) is thrown and the player losses k dollars if the top point k is odd, and gains 2k dollars if it is even. Find E(X), V(X) where X denotes the number appearing on the top.

Problem 1.26 If $p(x) = 6/[\pi^2 x^2]$ for $x = 1, 2,\ldots, \infty$, find $E(X^2)$. Does the variance of X exist?.

Problem 1.27 If $X \sim p(n) = 6H_n/[\pi^2 n(n+1)]$ where $H_n=1+1/2+1/3+\cdots 1/n$ is the n-th harmonic number, for $n = 1, 2, \ldots, \infty$, find $E(X^2)$. Does the variance exist?.

Problem 1.28 If $f(x, y) = x + y$ for $0 < x < 1, 0 < y < 1$, find $E(Y \mid X)$ and $E(X \mid Y)$.

Problem 1.29 Find the unknown constant K in $f(x) = K/(x + c)^{n+1}$, $x>0$, where n is an integer, and c is a real constant. Prove that all ordinary moments of order up to n–1 are non-existent for this distribution.

Problem 1.30 When is $Cov(X, Y) = -\mu_x \mu_y$?.

Problem 1.31 If $X \sim EXP(\lambda)$ find $E(\frac{\sqrt{X}+\sqrt{Y}}{2})^2$.

Problem 1.32 Prove that $M_{ax}(t) = M_x(at)$ and $M_{ax+b}(t) = e^{bt} M_x(at)$. Deduce that $M_{(x-\mu)/\sigma}(t) = e^{-\mu t/\sigma} M_x(t/\sigma)$.

Problem 1.33 Prove that the expected value of MD is $E(MD) = \delta\sqrt{1 - \frac{1}{n}}$ where n is the sample size and δ is the population MD.

Problem 1.34 If $f(x) = K \exp(-|x|)$ for $-\infty < x < \infty$, find K, E(X) and the MGF.

Problem 1.35 Show that $E(cY \mid X) = c\, E(Y \mid X)$ where c is a constant.

Problem 1.36 If $X \sim EXP(\lambda)$, find $E[\sqrt{X}]$.

Problem 1.37 If $f(x) = \binom{x-1}{r-1} p^r q^{x-r}$, $x = r, r + 1,\ldots$, find the MGF and derive the first two moments. What is the survival function?

Problem 1.38 Prove that $\sigma^2 = E(X(X-1)) + E(X) - [E(X)]^2$. Apply it to find the variance of Geometric and Poisson distributions.

Problem 1.39 For the Poisson distribution with parameter λ, prove that

$$Q(k, \lambda) = P(X \le k) = (1/\Gamma(k+1)) \int_{\lambda}^{\infty} e^{-x} x^k dx$$

satisfies $Q(k+1, \lambda) = (1 + \lambda/(k+1)) Q(k, \lambda) - \lambda/(k+1) Q(k-1, \lambda)$.

Problem 1.40 Prove that $E(X-k)^2 = \sigma^2 + [E(X) - k]^2$.

Problem 1.41 If X and Y are IID Bernoulli random variables defined on (0, 1) with Pr(X = 0) = p and Pr(Y = 0) = q, prove that Pr$[X + Y \ge 1] = 1-pq$.

Problem 1.42 For the geometric distribution f(x; p) = $q^x p$, prove that $E[\binom{x}{k}] = (q/p)^k$.

Problem 1.43 If X and Y are IID random variables with joint PDF f(x, y) = x + k y^2, for $0 \le x, y \le 1$, find (i) the constant K and (ii) distribution of X + Y.

Problem 1.44 A number x is chosen randomly from the sample space of an equally likely random experiment S = {1, 2, 3, 4, 5, 6, 7, 8}. What is the probability that Pr(x + 5/x)$\le$6)?.

Problem 1.45 If M(X, Y) = (X + Y)/2 is the mean of two positive random variables, when does equality hold in E[M(X, Y)] = M(E(X), E(Y))?

Problem 1.46 If G(X, Y) = $\sqrt{XY}$ is the geometric-mean of two positive random variables, prove that E[G(X, Y)]$\le$ G(E(X), E(Y)). If H(X,Y) = 2/[1/X + 1/Y] is the harmonic-mean of two random variables, prove that E[H(X, Y)]$\le$ H(E(X), E(Y)).

Problem 1.47 If X and Y are two positive random variables, prove that $E(XY/[X+Y]) \le E(X)E(Y)/E(X+Y)$.

Problem 1.48 Find $E((1 + \frac{x}{n})^n)$ for the binomial distribution with n trials where x is the number of successes. What is the limit of this expectation as n $\to \infty$?.

Problem 1.49 Find the MGF of logarithmic series distribution f(x) = $c\theta^x/x$, x = 1, 2, $\cdots$ where c = $-1/\log(1-\theta)$, and 0< θ <1. Prove that rth factorial moment $\mu_{(r)} = c(r-1)![\theta(1-\theta)]^r$.

Problem 1.50 If X $\sim$ BETA(a, b), prove that E(X) = a/(a + b).

Problem 1.51 If $X \sim$ BETA(a, b–a), prove that $M_x(t) =_1 F_1(a, b; x) = \frac{\Gamma(b)}{\Gamma(a)\Gamma(b-a)}$ $\int_0^1 e^{yt} t^{a-1}(1-t)^{b-a-1} dt$ is the confluent hypergeometric function. Hence show that E(X) = a/b.

Problem 1.52 If f(x) = $1/\pi$ for $0 < x < \pi$, show that E[sin(x)] = $2/\pi$.

Problem 1.53 If X > 0, find constants a, b, c such that $E[X-t|X > t] = c\int_a^b [1 - F(x)]dx$.

Problem 1.54 Prove that $E[X^2] = \sum_{k=0}^{\infty}(2k + 1)P[X > k]$.

Problem 1.55 What is the expected value of an indicator variable?

Problem 1.56 If $X \sim$ EXP(λ) find $E[e^{cx}]$ where c is constant.

Problem 1.57 If X has Geometric distribution, find $E[(-1)^x]$.

Problem 1.58 If $X \sim$ BINO(n, p) prove that $E(n-X)^2 = nq[n-p(n-1)]$.

Problem 1.59 If $X \sim$ BINO(n, p) find $E(\frac{x-1}{x+1})$.

Problem 1.60 The PMF of a discrete random variable is given by f(x) = K(|x| + 1) for x = –3, –2, –1, 0, 1, 2, 3. Find K and the CDF. Evaluate F(2) and Pr[X ≥ 0].

Problem 1.61 If f(x) = Kexp($-|x|/(\sigma/\sqrt{2})$) is a PDF, find K, MGF and E(X).

Problem 1.62 If f(x) = $K/2^{x-1}$ for x = 1, 2, 3, 4; find K, E[X] and Pr[X ≥ 2].

Problem 1.63 If f(x) = $(1 - \frac{\mu}{n})^n \binom{n}{x}(\frac{\mu}{n-\mu})^x$, x = 0, 1, . . .n, find the mean and variance.

Problem 1.64 If f(x, y) = C(x + y) for $0 < x < y < 1$, find C. What is the value of $E(X^2 + Y^2)$?

Problem 1.65 If $\phi(x)$ is a real-valued, monotonic function of a positive random variable X, prove that $E[\phi(x)] = \phi(0) + \int_0^{\infty}[1-F(x)]\, \partial\phi(x)/\partial x$. Hence derive that $E[X^n] = n\int_0^{\infty} x^{n-1}[1-F(x)]\, dx$.

Problem 1.66 What are the conditions for a function to be a moment generating function? Are the following functions true MGF?
(a) $e^{a(t-1)+b(t-1)^2}$, (b) $e^{a(t-1)/(1-bt)}$, (c) $e^{a(t-1)+b(t^2-1)}$, (d) $e^{|(t-1)/(t^2-1)|}$.

Problem 1.67 Suppose an urn contains m red balls and n blue balls. If r balls are drawn with replacement, what is expected number of blue balls drawn?.

Problem 1.68 If X and Y are IID distributed as lognormal (μ, σ^2), (i) find E[XY] (ii) approx. value of E[XY log(XY)].

Problem 1.69 If X is a discrete symmetric random variable taking only two values (–k, +k) with probability (P[X = k] = P[X = –k]), find expected value of $\sin(\pi X)$.

Problem 1.70 Prove that $COV(\overline{X}, X_i - \overline{X}) = 0$ for any random sample.

Problem 1.71 Prove that $COV(\overline{X}, X_i) = \sigma^2/n$ for any random sample.

Problem 1.72 Use $x^2 = x(x-1) + x$ and $x^3 = x(x-1)(x-2) + 3x(x-1) + x$ to find the higher-order expected values of a Poisson random variable. What is the expression to find $E[x^4]$?.

Problem 1.73 Find the mean and variance of the distributions:– (i) f(x, n) = (n/2) sin(nx), $0 \le x \le \pi/n$, n>0 is real (ii) f(x, n) = (n/2) cos(nx), $-\pi/2n \le x \le \pi/2n$, n > 0 is real.

Problem 1.74 Prove that $\sigma^2_{Y|X} = \sum_k y_k^2 f(y|x) - \mu^2_{Y|X}$.

Problem 1.75 If X is continuous, prove that $\mu'_r = \int_0^\infty r x^{r-1}[1 - F_X(x) + (-1)^r F_X(-x)]$ dx = $E(X^r)$.

Problem 1.76 Find the expected values of x^2 and x^3 in a random experiment of tossing a fair die, where X is the number shown at top.

Problem 1.77 If X and Y are IID GEO(p) and GEO(q) distributions, prove that E[X| X≤Y] = 1/(p + q – pq).

Problem 1.78 If X and Y are IID GEO(p) and GEO(q) distributions, find Pr[X > Y].

Problem 1.79 Prove that the factorial moments for the following distributions are as given:
BINO(n, p): $E[X(X – 1)..(X – r + 1)] = n_{(r)} p^r$.
HYPG(N, n, p): $n_{(r)} N p_{(r)}/N_{(r)}$ if $f(x) = \binom{Np}{k}\binom{Nq}{n-k}/\binom{N}{n}$
GEOM(p): $r! \, q^{r-1}/p^r$
NBINO(n, p): $r(r + 1)..(r + s) \, (q/p)^s$.

Problem 1.80 If $X_1, X_2, \ldots, X_n$ are IID random variables with the same mean μ, and same σ^2, find the expected value and variance of the arithmetic mean of X's.

Problem 1.81 If X is a standard normal random variable, and $Y = \exp(X^2)$ find E(Y).

Problem 1.82 If X is Binomial(n, p), where n is even and $0 < p < 1$, for what value of p is $\Pr[x = n/2]$ maximum?.

Problem 1.83 If X is Poisson(λ), for what value of λ is the $\Pr[x = 2]$ maximum?

Problem 1.84 If the CDF of Levy distribution (which is the distribution of $U = 1/Z^2$) is given by $G(u) = 2[1 - \Phi(1/\sqrt{u})]$ for $u \in [0, \infty)$, prove that the quantile function is $Q(p) = 1/[\Phi^{-1}(1 - p/2)]^2$ for $p \in [0, 1)$.

Problem 1.85 If X is a real-valued continuous random variable, prove that $E[X^2] = 2\int_0^\infty x[1 - F(x) + F(-x)]dx$.

Problem 1.86 If X and Y are two IID exponential random variables with parameters λ and δ, prove that the mean of $Z = \min(X, Y)$ is $1/(\lambda + \delta)$.

Problem 1.87 The steady-state waveform of magnetic flux when the inductor voltage is square-wave form at core saturation is given by

$$f(x; V, f, T) = \begin{vmatrix} V[x - \frac{1}{4f}] & 0 \le x \le T/2 \\ -V[x - T/2] & T/2 < x \le T \end{vmatrix}$$

Find the mean and variance.

Problem 1.88 The mean-excess function of a variate is defined as $\hat{E}[X] = (1/S(x))\sum_u^\infty (1 - -F(u))$ if X is discrete, and $\hat{E}[X] = (1/S(x))\int_u^\infty (1 - F(u))du$ if X is continuous, where $S(x) = 1 - F(x)$ is the survival function. Find $\hat{E}[X]$ of Poisson and exponential variates.

Problem 1.89 When a cell phone is powered on, it is registered with a base station. Each base station has a "cell" which is the coverage region (say a circular or square region) around it. When the caller moves from place to place, they may move out of one region and into an adjacent region. The phone company automatically detects it and "hands over" the phone identity to the new base station. A phone company has noticed that the majority of subscribers do not change their base station during their call, but the proportion of subscribers who change their base station is an upper truncated Poisson distribution with $\lambda = 0.04$, and truncation point 4. Find the expected percentage of subscribers who change their base station, and $\Pr[X \ge 2]$ where X denotes the number of hand overs.

Problem 1.90 If the CDF of a continuous random variable is F(x)=

$$
\begin{cases}
0 & \text{if } x < 2 \\
c(\frac{x^2}{2} - 2(x-1)) & \text{if } 2 \le x < 4 \\
c(\frac{-x^2}{2} + 2(3x-7)) & \text{if } 4 \le x < 5 \\
F(x) = 1 & \text{for } x \ge 5.
\end{cases}
$$

find the PDF and the mean.

Problem 1.91 If the PMF of a discrete random variable is

$$
f(x) = \begin{cases}
0.25 & \text{if } x = -1 \\
0.50 & \text{if } x = 0 \\
0.25 & \text{if } x = +1
\end{cases}
$$

how is the mean related to $P[X = 1]$.

Problem 1.92 If the CDF of a continuous random variable is

$$
F(x) = \begin{cases}
0 & \text{if } x < 0 \\
c\frac{x^2}{2} & \text{if } 0 \le x < 1 \\
c(2x - 1 - x^2) & \text{if } 1 \le x < 2 \\
F(x) = 1 & \text{for } x \ge 2.
\end{cases}
$$

find the PDF and the mean.

Problem 1.93 If the PMF of a discrete random variable is

$$
f(x) = \begin{cases}
\exp(-\lambda)\, p^x \sum_{j=1}^{x} \binom{x-1}{j-1}(\lambda q/p)^j / j! & \text{for } x = 1, 2, \dots \\
\exp(-\lambda) & \text{for } x = 0 \\
0 & \text{elsewhere}
\end{cases}
$$

prove that the mean is λ/q. Find the PGF.

Problem 1.94 If the CDF of a continuous random variable is

$$
f(\theta) = \begin{cases}
0 & \text{if } \theta < 0, \\
\tan(\theta)/2 & \text{if } 0 \le \theta \le \pi/4, \\
1 - \tan(\pi/2 - \theta)/2 & \text{if } \pi/4 \le \theta \le \pi/2, \\
1 & \text{elsewhere}
\end{cases}
$$

prove that the mean is $\pi/4$. Find the PGF.

Problem 1.95 Prove that the memory-less property of exponential distribution is equivalent to $G(u + v) = G(u)*G(v)\ \forall u, v > 0$ where $G(u) = \Pr[X > u]$.

Problem 1.96 If both X and Y are independent gamma distributed, prove that (i) $E(Y|X) = cX + b$, (ii) $Var(Y|X) = b$, (iii) $E((Y–X)^2|X) = b$.

Problem 1.97 If $f(x; n, \mu) = \binom{n}{x}(1 - \mu/n)^n(\mu/(n - \mu))^x$ where $x = 0, 1, \ldots, n$ prove that $E(X) = \mu$ and $Var(X) = \mu(1 - \mu/n)$.

Problem 1.98 If X, Y are independent normal random variables with the same variance, find $E[(X + Y)^4|(X - Y)]$.

Problem 1.99 Find the mean and median of the weighted exponential distribution (WED) with PDF $f(x; c, m) = m(c + 1)/c \exp(-mx)(1-\exp(-mcx))$, for $x > 0$. If X and Y are IID WED, find the distribution of X + Y.

Problem 1.100 Suppose you toss a fair die once and note down the number N that shows up $(1 \leq N \leq 6)$. You then toss a fair coin N times. Let X denote the number of heads that you get in N tosses of the coin. Find E(X) and V(X).

Problem 1.101 If X and Y are independent random variables, prove that $P(Y \leq X) = \int_{-\infty}^{\infty} F_Y(x)f_X(x) \, dx = 1 - \int_{-\infty}^{\infty} F_X(y)f_Y(y)dy$.

Problem 1.102 Prove that the population variance can be expressed as the values of CDF $(F_X(x))$ and its first two derivatives evaluated at $x = 1$.

Problem 1.103 Prove that the population variance can be expressed using the second derivative of $K_x(t)$ as $K_x''(x)$ evaluated at $x = 0$.

Problem 1.104 A size-biased Maxwell distribution has PDF $f(x;a) = K(1 + cx) x \exp(-x^2/(2a^2))$. Prove that $1/K = (1 + a) a^2$ and find the mean.

Problem 1.105 A telephone carrier notices that the average duration of cell-phone calls among teenagers is distributed as a left-truncated exponential distribution with $\lambda = 1/2400$ s and truncation point 20 s. What is the expected percentage of phone calls that take more than 5 min? What is the variance of duration of all phone calls?

Problem 1.106 A site offers HTTP and FTP connections. Number of new customers who connect to HTTP server is Poisson distributed with $\lambda = 20$ for a time interval of 1 min. On the FTP server is Poisson distributed with $\lambda = 3$ for same time period. If both events are independent, what is the expected number of customers connecting to the site in 4 min?.

Problem 1.107 The number of empty taxicabs that arrive at a city center between 8 and 10 AM is Poisson distributed with mean 5 in 3 min. What is the expected number of minutes a person has to wait if there are no others in the queue? What is the expected number of cabs that arrive in 72 s?

Problem 1.108 If X is CUNI(a, b) find the distribution and expected value of Y = (2X–(a + b))/(b – a).

Problem 1.109 If $F(x, y) = \Pr[X \leq x, Y \leq y]$, prove that $\lim_{x \to \infty} F(x, y) = F(y)$.

Problem 1.110 If $F(x, y) = C - \exp(-x) - \exp(-y) + \exp(-(x + y))$ is a joint CDF, (i) find the unknown C, (ii) prove that $E(X) = E(Y)$, (iii) check if X and Y are independent.

Problem 1.111 If X and Y are two symmetric random variables with joint CDF F(x, y) = F(y, x), prove that the marginal distributions of X and Y are the same.

Problem 1.112 High-rise structures at earth-quake-prone areas are designed to withstand powerful earthquakes. From past data, it is found that the probability of an earthquake in a year is 0.091, and the probability of a building collapse after the earthquake is 0.004. The cost of constructing a high-rise building is C_0 and the cost of repair after damage is C_r. If a building portfolio comprises of n(>2) buildings in a city neighborhood, find the expected value of the cost incurred in 10 years (i) if no information on the earthquakes are available (ii) if it is assumed that at least 2 earthquakes are likely to occur.

Problem 1.113 An automated robot controlled inventory warehouse has racks of length 120 m on both sides of an alley. The robot is equally likely to break down anywhere on the stretch of 120 m. Where should a spare robot be located so that it can immediately take over the task of the broken down robot in minimal waste of time?.

Problem 1.114 Check whether a function defined as $f(x; c) = 2c \, x \, \exp(-cx^2)$ over $[0, \infty)$ is a PDF.

Problem 1.115 If $X_i \sim$ IID BETA-II (a, b_i), find expected value of harmonic mean.

Problem 1.116 The size-biased unit exponential distribution has PDF $f(x) = xe^{-x}$, for x > 0. Find the CDF and the mean.

Problem 1.117 The size-biased unit Laplace distribution has PDF $f(x) = x \exp(-|x|)$, for $-\infty < x < \infty$. Find the CDF and the mean.

Problem 1.118 If f(x, y) = a(x–y) where $0 < x < b$, $-x < y < x$, find the unknown in terms of b, and the mean.

Problem 1.119 An auto-emission test center has found that on the average one in 8 automobiles fail in the emission test, and needs tune-up. The distribution of tune-up time in hours is EXP(2.5). If 100 vehicles are tested per month, find the expected number of hours spent on servicing of failed vehicles.

Problem 1.120 A cereals manufacturer offers a promotional coupon with a new brand of cereal pack. Two types of coupons (that carry either 1 point or 2 points) are printed, and either of them is put in selected packs so that some packs do not contain a coupon. Probability that a customer will find a 1-point coupon is p, and a 2 points coupon is q. If a customer purchases n packs of the cereal, what is the expected number of points earned?.

Problem 1.121 If X and Y are two independent random variables, and U = XY, verify whether Var(U) = Var(X) μ_y^2 + Var(Y) μ_x^2 + Var(X) Var(Y), where μ_x = E(X) and μ_y = E(Y).

Problem 1.122 Prove that for the central chi-square distribution E(1/X) = 1/(n–2).

Problem 1.123 Prove $K_{aX+b}(t) = K_X(at) + bt$ where K denotes cumulant generating function.

Problem 1.124 If g(x) is a convex function, prove that $E[g(x)] \geq g(E[x])$ provided that $E[|g(x)|] < \infty$.

Problem 1.125 Prove using mathematical expectation that Cov(X, Y) $\leq [V(X)V(Y)]^{1/2}$ where V() denotes variance and Cov() is the covariance. When does equality hold?.

Problem 1.126 A random variable X is called regularly varying if $\lim_{x \to \infty} \Pr[|X| > xt]/\Pr[|X| > x] = t^{-c}$ for c > 0. Check whether the geometric and Pareto laws are regularly varying.

Problem 1.127 If $X_1, X_2, \ldots, X_N$ are IID and $S_N = X_1 + X_2 + \cdots + X_N$, prove that $E(\sum_{i=1}^{N} X_i) = E(N)E(X)$, and $M_{S_n}(t) = \prod_{i=1}^{n} M_{x_i}(t)$.

Problem 1.128 If $X_1, X_2, \ldots, X_N$ are IID each with the same mean μ and same variance σ^2, find the second moment and variance of a random sum $S_N = X_1 + X_2 + \cdots + X_N$.

Problem 1.129 What is the limiting value of $\lim h \downarrow 0 (E[e^{hx} - 1])/h$.

Problem 1.130 If X~ $N(\mu, \sigma^2)$, prove that $E[\Phi(x)] = \Phi(\mu/\sqrt{1+\sigma^2})$ where $\Phi()$ denotes the CDF of a standard normal distribution.

Problem 1.131 An audio signal S is corrupted with background noise B. If S is uniformly distributed in the range –c to +c, but the noise B is uniformly distributed in the range 0–2d where d < c, what is the expected value of signal plus noise? What is the covariance COV(S, B) assuming that signal and noise are coming from independent sources?.

Problem 1.132 Suppose two fair dice are tossed. Find the density function of (X_1, X_2) where X_1 and X_2 are the scores that show up.

Problem 1.133 Three random variables U, V, Y are defined as follows: U = min$\{X_1, X_2\}$, the minimum score, V = max$\{X_1, X_2\}$, the maximum score, Y = X_1+ X_2, the sum of the scores. Find E(U), E(V), E(Y), E(Y|X_1).

Problem 1.134 If X is a continuous random variable with PDF f(x,c) = $(c-1)/x^c$ for x>1, and zero otherwise, what is the value of c if E(X) = 2.

Problem 1.135 A continuous random variable defined on the interval [1,2] has CDF F(x) = $C(x^2 - 2x + 2)$. Find C, the mean and variance.

Problem 1.136 An insurance company pays a policyholder a fixed amount M per day for the first 3 days of hospitalization and M/2 for a maximum of 2 more days. If the probability that a patient is likely to be hospitalized for x days is Poisson(0.40) distributed, find the expected payment for 5 days.

References

Chakraborti, S., Jardim, F. & Epprecht, E. (2019). Higher-order moments using the survival function: The alternative expectation formula, *The Amer. Statn.*, 73(2), 191–194. https://doi.org/10.1080/00031305.2017.1356374

Chattamvelli, R. (1995). A note on the noncentral beta distribution function, *The Amer. statn.*, 49, 231–234, https://doi.org/10.2307/2684647

Chattamvelli, R. (2016). *Data Mining Methods*, Alpha Science, Oxford, UK.

Chattamvelli, R. & Jones, M.C. (1995). Recurrence relations for noncentral density, distribution functions, and inverse moments, *J. Stat. Compu. and Simu.*, 52(3):289–299. https://doi.org/10.1080/00949659508811679

Chattamvelli, R. & Shanmugam, R. (1998). Computing the noncentral beta distribution function, Algorithm AS-310, *Appl. Stat.*, Royal Stat. Soc., 41:146–156. https://doi.org/10.1016/0167-9473(94)90162-7

Chattamvelli, R. & Shanmugam, R. (2019). *Generating Functions in Engineering and the Applied Sciences*, Springer. https://link.springer.com/book/10.1007/978-3-031-21143-0

Chattamvelli, R. & Shanmugam, R. (2021) Continuous Distributions in Engineering and the Applied Sciences – Part 1, Springer. https://link.springer.com/book/10.1007/978-3-031-02430-6

Chattamvelli, R. (2024). *Correlation in engineering and the applied sciences: Applications in R*, Springer. https://link.springer.com/book/9783031510144

Dezfouli A.R, Khonsari M.N., et al. (2023). Waist to height ratio as a simple tool for predicting mortality: a systematic review and meta-analysis. *Intl. J. Obes.* (Lond). 47(12):1286–1301, Epub 2023 Sep 28. PMID: 37770574. https://doi.org/10.1038/s41366-023-01388-0

Giri, N.C. (2004). Approximations and tables of beta distributions, in *Handbook of beta distribution,* Gupta, A.K., Nadaraja, S. (eds), Marcel Dekker. https://doi.org/10.1201/9781482276596

Jiang, X. & Nadarajah,S.(2019). On characteristic functions of products of two random variables, *Wireless Personal Commu.*, https://doi.org/10.1007/s11277-019-06462-3

Jones, M.C. (2002). The complementary beta distribution, *J. stat. plan. & inf.*, 104, 329–337. https://doi.org/10.1016/S0378-3758(01)00260-9

Jones, M.C., Marchand E. & Strawderman, W.E. (2018). On an intriguing distributional identity, *The Amer. Statn.*, 73(1) 16–21, https://doi.org/10.1080/00031305.2017.1375984

Raja Rao, B(1986). A note on the moments of non-central chi-squared and F distributions, *Metron* 44, 121–130,

Whittle, P.(2000). *Probability via expectation*, 4th ed., Springer, https://link.springer.com/book/10.1007/978-1-4612-0509-8

Wilf, H.S. (1994). *Generatingfunctionology*,3rd edn, Academic press, NY. https://doi.org/10.1201/b10576

Functions of Random Variables

2

This chapter discusses single random variables and its transforms. Various types of transformations such as sum, squares, square-roots, absolute-value, reciprocals, transcendental and arbitrary functions are discussed and illustrated with numerous examples. Distribution of minimum and maximum, integer and fractional parts, sum of squares of IID RVs as well as the distribution of CDF and its inverse are also described. A summary table of transformations is provided to find the transformed densities easily. These results are used to express the mean deviation of continuous random variables as a simple integral from lower limit to F(mean) where F() is the CDF. This chapter ends with a discussion of applications of functions of random variables in various fields.

2.1   Introduction

This chapter discusses the distribution of functions of a single random variable (RV). There are many situations where simple functions of RVs have well-known distributions. One example is the relation between a standard normal and a chi-square distribution. As shown below, the square of a standard normal is chi-square distributed with one degree of freedom (DoF). If there are several independent standard normal variates, the sum of their squares is also chi-square distributed with n DoF where n is the number of variates.

The statistical analysis of complex relationships is greatly simplified by variate transformations. For example, several non-normal RVs are involved in most practical applications. A common technique is to convert them into equivalent (or approximate) standard distributions (e.g.: N(0, 1)) using transformation methods such as Orthogonal transforms, Rosenblatt transform or Nataf transform. These are discussed at the end of the chapter.

2.2   Distribution of Functions of RVs

The classical method known as *the method of distribution function* (MoDF) is useful when the CDF has closed form. The CDF of the transformed variable is easily obtained using MoDF when the transformation is strictly increasing or decreasing function, from which the PDF follows readily by differentiation (see Table 2.3) in the continuous case. This method can be used to find the PDF of the logistic, Gumbel, Pareto and uniform distributions. There are many ways to derive such related distributions. The most popular among them are the (i) CDF method (ii) MGF (or ChF) method (iii) trigonometric transformations (iv) geometric reasoning (v) using Jacobians and (vi) copula based methods (Chap. 3).

2.2.1   Distribution of Translations

These are obtained by a change of origin transformation $Y = X \pm c$. As the variate values are shifted either to the right (c is positive) or to the left (c is negative), the PDF or PMF remains the same, but the range is modified accordingly. A special translation is called reflection as $Y = -X + c$ where c is a location measure. If X is symmetric, $Y = -X + c$ results in the same distribution if c is the mean, median or mode (all of which coincides for symmetric laws). This transformation is usually applied along with one of the following transformations such as the change of scale or square transformation.

Translations of CUNI(a, b) Distribution

If X is CUNI(a, b) find the distribution of (i) $Y = X–a$ (ii) $Y = X–(a + b)/2$.

Solution As both of these are change of origin transformations, the PDF remains the same as $f(y) = 1/(b–a)$. In case (i) the lower-limit becomes 0 and upper limit becomes (b–a), so that $f(y) = 1/(b–a)$, $0 < y < (b–a)$. In case (ii) the lower-limit is $a–(a + b)/2 = (a–b)/2 = –(b–a)/2$, and the upper limit is $b–(a + b)/2 = (b–a)/2$. This gives $f(y) = 1/(b–a)$, $–(b–a)/2 < y < (b–a)/2$. As $(a + b)/2$ is the mean of a rectangular distribution, this transformation is a reflection.

Transformation of binomial

If X is BINO(n, p) find distribution of (i) Y = 1–X/n, (ii) Y = n–X.

Solution 1–X/n takes the fractional values {1, (n–1)/n, (n–2)/n, ..., 1/n, 0} with corresponding binomial probabilities q^n, $\binom{n}{1}q^{n-1}p,...,p^n$. (ii) Y = n–X is distributed as BINO(n, q) where the probability of success is q = 1–p with PDF f(x) = $\binom{n}{x}q^x p^{n-x}$.

Definition 2.1 If f(x) and g(y) are two continuous functions, then h(x) = α f(x) + (1–α) g(y) where $0 \le \alpha \le 1$ is called a convex combination of f(x) and g(y).

The range of h(x) is from the minimum of the ranges of (f(x), g(y)) to the maximum of the ranges of (f(x), g(y)).

Convex Combination of RVs

If F and G are two distributions symmetric around zero, with unit second moment, prove that H = c F + (1–c) G is identically distributed for $0 < c < 1$.

Solution Given F(–x) = 1–F(x), G(–x) = 1–G(x), and c > 0 so that 1–c > 0, $\mu = 0$ (as both are symmetric around 0). Take variance of both sides and use Var(cX) = c^2 Var(X) to get Var(H) = c^2 Var(F) + $(1 - c)^2$ Var(G). Using the given condition that both of them have unit second moment, the RHS is obviously positive. Hence by the central limit theorem, the normed convex combination is identically distributed.

2.2.2   Distribution of Constant Multiples

First consider the case of a discrete RV X. If c is an integer, then Y = c*X is a change of scale transformation that maps X values to positive or negative numbers. Hence depending on the sign of c, Y could belong to the same family of distribution. As an example, if X is POIS(λ) with MGF exp($\lambda(e^t - 1)$), Y has MGF exp($\lambda(e^{ct} - 1)$). This is the MGF of a Poisson distribution. If c is a fraction, the distribution is still well-defined if we assume that X takes fractional values (but still it is discrete distribution due to the discontinuity). See Sect. 2.3.1 below. Distribution of constant multiples in the continuous case can be found using the MoDF method (described below) or using Jacobians. However, some of the continuous distributions (e.g.: power-law distribution) are scale invariant and affects only the constant multiplier.

2.2.3 Method of Distribution Functions (MoDF)

The MoDF is a simple and powerful method to find the PDF of a variety of continuous transformations. Consider the general transformation $Y = h(X)$. The MoDF works when (i) h(x) is either an increasing or a decreasing function of x without discontinuities, (ii) the first derivative of h(x) exists throughout the range of the variate, (iii) h(x) is invertible (so that $x = h^{-1}(y)$, is uniquely solvable) (iv) F(x) is differentiable once. We illustrate the use of MoDF for various forms of h(x) in their respective sections. Consider the transformation Y $= h(x) = c*X + b$ where X has a known distribution. It satisfies all the 3 conditions on h(x). The CDF of Y is

$$G(y) = \Pr(Y \leq y) = \Pr(cX + b \leq y) = \Pr(X \leq (y - b)/c) = F((y - b)/c). \qquad (2.1)$$

Note that in (2.1), G(.) is the CDF of Y, and F(.) is the CDF of X. As the CDF of Y contains only y; and the constants (b, c), we have simply expressed y in terms of x. Differentiate wrt y to get

$$g(y) = (\partial/\partial y) F((y - b)/c) = (1/c) * f((y - b)/c). \qquad (2.2)$$

Here we need to consider two cases as $c > 0$ or $c < 0$. The above result holds for $c > 0$. When $c < 0$ (Y $= h(X)$ is a decreasing function), we get

$$G_Y(y) = \Pr(cX + b \leq y) = \Pr(x \geq (y-b)/c) = 1 - F((y-b)/c). \qquad (2.3)$$

Differentiation gives us $g(y) = (-\partial/\partial y)F((y - b)/c) = (-1/c) * f((y - b)/c)$. Combine both cases to get

$$g(y) = (\pm \partial/\partial y) F((y-b)/c) = (1/|c|) * f((y - b)/c) \qquad (2.4)$$

where the vertical line means "absolute value" which absorbs the $\pm$ sign. Thus the PDF of Y is easily obtained from that of X. The only assumption made here is that F(y) is once differentiable, as the other conditions are satisfied by the linear transformation. The constant c can be any nonzero real number. Depending on whether $|c| < 1$ or > 1, the new range is either expanded or contracted.

Constant Multiple of CUNI(a, b) Distribution

If X is CUNI(a, b) find the distribution of Y $= (2X-(a + b))/(b - a)$.

Solution Write Y $= (2X-(a + b))/(b - a)$ as $c*X + d$ where $c = 2/(b-a)$ and $d = -(a + b)/(b - a)$. Solve for X to get $x = (Y-d)/c$. Then use (2.2) to get $g(y) = f((y-d)/c)/c$. As X is CUNI(a, b), $f(x) = 1/(b-a)$, $a < x < b$. As this does not involve x, putting $x = (y-d)/c$ has no effect. The range is modified as -1 and $+1$. Substitute the values of c and d to get $g(y) =$

$(1/(b–a))/(2/(b–a)) = 1/2, -1 < y < 1$. If $X \sim \text{CUNI}(0, b)$ and $Y \sim b–X$, it is trivial to find the distribution of X–Y because $X–Y = 2X–b$ which is $\text{CUNI}(–b, b)$ using above result.

Scaled exponential

If X is $\text{EXP}(\lambda)$, find the distribution of $Y = cX$.

Solution This is easy using the CDF method. The CDF of Y is $F(y) = P(cx \leq y) = P(x \leq y/c) = 1 - \exp(-\lambda y/c)$. Differentiate wrt y to get the PDF as $f(y) = (\lambda/c)\exp(-\lambda y/c)$, for $y > 0$.

Difference of rectangular distributions

If X and Y are IID rectangular distributed in [0, b], find the distribution of X–Y.

Solution Let $U = X–Y$. The CDF of U is $F(u) = \Pr[U \leq u] = \Pr[X–Y \leq u]$. As the joint PDF of (x, y) is $1/b^2$, the CDF of U is $\frac{1}{b^2} \int \int_{X-Y \leq u} dxdy$. We have to consider two cases, $-b \leq X–Y \leq 0$ and $0 < X–Y \leq b$. In the first case the CDF is $F(u) = (1/b^2) \int_0^{b+u} \int_{x-u}^{b} dydx = (1/b^2) u^2/2 + u + 1/2$. In the second case $F(u) = 1-(1/b^2) \int_u^b \int_0^{x-u} dydx = (1/b^2) - u^2/2 + u + 1/2$. Differentiate wrt u to get the PDF as

$$f(u) = \begin{cases} (u+1)/b^2, & -b < u < 0; \\ (2-u)/b^2, & 0 \leq u < b. \end{cases}$$

See Chap. 3, Sect. 3.3.1 for an alternate derivation.

2.2.4   Distribution of Absolute Value (|X|) Using MoDF

The MoDF can easily be adapted to find the distribution of $Y = |X|$. This is meaningful only when X takes both positive and negative values (if X assumes only negative values, then $|x| = -x$). As above, let G(y) be the CDF of Y. Then

$$G(y) = \Pr(Y \leq y) = \Pr(|X| \leq y) = \Pr(-y \leq X \leq y) = F(y) - F(-y). \qquad (2.5)$$

Differentiate both sides to get

$$g(y) = f(y) + f(-y), y > 0. \qquad (2.6)$$

In the particular case when Y is symmetric, f(y) = f(−y) so that g(y) = 2 f(y).

Absolute value of CUNI(−π/2, π/2)

If X is CUNI(−π/2, π/2) variate, find the distribution of Y = |X|.

Solution As the CUNI(−π/2, π/2) distribution exhibits special symmetry around zero point, f(−x) = f(x). We know f(x) = 1/π for −π/2 ≤ x ≤ π/2. Thus using (2.5) the PDF of Y = |X| is f(y) = 2*f(x) = 2/π for 0≤ x ≤ π/2.

Distribution of absolute value of Cauchy Variate

If X is a standard Cauchy variate, find the distribution of Y = |X|.

Solution We know that f(x) = 1/[π(1 + x²)], which is symmetric in x. Using (2.5), the distribution of Y is 2*f(x) = 2/[π(1 + y²)] for 0≤ y < ∞.

2.2.5 Distribution of F(x) Using MoDF

Let F(x) be the CDF and $F^{-1}(x)$ be the inverse CDF of a continuous RV. Obviously the minimum value F(x) can take is 0, and the maximum value is 1. In addition, F(x) is always an increasing function of X.

Distribution of F(x)

If X is a continuous variate, U = F(x) is uniformly distributed in [0, 1]. Consider

$$F(u) = \Pr(U \leq u) = \Pr(F(x) \leq u) = \Pr(x \leq F^{-1}(u)) = F[F^{-1}(u)] = u. \qquad (2.7)$$

The CDF of a rectangular distribution f(x) = 1/(b−a) is (x−a)/(b−a). Put a = 0, b = 1 to get F(x) = x. Equation (2.7) then shows that U has a unit rectangular distribution U(0, 1). In essence, this states that any continuous RV can be transformed to a uniform distribution (this fact is used in copula models Sect. 3.7). A natural extension of this result is to transform one RV (say X) to another (say Y) where neither X nor Y are uniform distributed. We could do it in two steps by first transforming X to U and then from U to Y. This can be done in a single step when the CDF of Y is invertible. Let F(x) and F(y) denote the respective CDFs. Equate both to get F(x) = F(y) ∀ x. As Y is invertible, operate by F_y^{-1} on both sides to get $F_y^{-1}F(x) = F_y^{-1}$ F(y) = y. The RVs are symmetrically transformable when both F(x) and F(y) are invertible. This technique can be extended to bivariate RVs and higher dimensional RVs.

2.2.6   Distribution of $F^{-1}(x)$

Distribution of $Y = F^{-1}(x)$ is well   tractable in the continuous case when x is defined on unit interval. If we define $F^{-1}(x)$ as the minimum value of y satisfying F(y)≥x, we could use the MoDF to find the distribution in certain cases. Obviously, $F^{-1}(x)$ is nondecreasing and satisfy (i) $F^{-1}(F(x)) \le$ x, for $-\infty < x < \infty$, and (ii) $F(F^{-1}(y)) \ge$ y for 0< y <1. Let $Y = F^{-1}(x)$. If there are no discontinuities,

$$G(y) = \Pr(Y \le y) = \Pr(F^{-1}(x) \le y) = \Pr(x \le F(y)) = F_x[F_x(y)] \qquad (2.8)$$

where we have used $F(F^{-1}(x)) = x$ (strictly). In the particular case when Y is U[0, 1], we have that F(y) = y. Substitute in the last term (in square bracket) of (2.8) to get the CDF of $F^{-1}(x)$ as G(y) = F(F(y)) = F(y) = y showing that Y is uniformly distributed with unit range (i.e. Y~U(0, 1)). Next consider the general case. As the derivative of F(F(y)) is unambiguously determined when the argument (of outer F()) ranges over the unit interval, we could differentiate both sides of (2.8) to get

$$g(y) = (\partial/\partial y) \, F[F(y)] = (\partial/\partial F(y)) \, F[F(y)] * \partial(F(y)/\partial y) = f[F(y)] * f(y) \qquad (2.9)$$

where we have used the function-of-a-function rule of differentiation because the inner F(y) being the CDF satisfies $0 \le F(y) \le 1$. An application of the above result is to find the mean deviation of continuous distributions given below.

Theorem 2.1 *If $f(F^{-1}(x))$ of a continuous distribution has closed form, and is integrable in the proper range, the mean deviation (MD) is given by*

$$MD = 2 \int_{t=0}^{F(\mu)} t \, dt / f(F^{-1}(t)) = 2 \int_{t=S(\mu)}^{1} (1 - t) dt / f(S^{-1}(t)). \qquad (2.10)$$

Proof The proof follows easily by putting y = $F^{-1}(x)$ and using properties of distribution functions in the above result. This can easily be extended to bivariate and multivariate distributions. As $F(\mu) = S(\mu) = 1/2$ for symmetric laws, this becomes

$$MD = 2 \int_{t=0}^{1/2} t \, dt / f(F^{-1}(t)) = 2 \int_{t=1/2}^{1} (1-t) \, dt / f(S^{-1}(t)), \qquad (2.11)$$

for symmetric distributions. This is illustrated below.

Mean deviation of Exponential Distribution

If X is EXP(λ) find the mean deviation.

Solution Consider the exponential distribution $f(x) = \lambda e^{-\lambda x}$ with mean $\mu = 1/\lambda$ and CDF $1-e^{-\lambda x}$. Put $x = 1/\lambda$ in the CDF to get $F(\mu) = 1-e^{-\lambda(1/\lambda)} = 1 - e^{-1} = (e-1)/e$. The inverse CDF is $F^{-1}(x) = -(\log(1 - x))/\lambda$ where log is to the base e (i.e. $-\ln(1-x)/\lambda$). Put the values in (2.10) to get the simple integral

$$MD = 2 \int_{t=0}^{(e-1)/e} t \, dt / \lambda(1 - t) \tag{2.12}$$

where we have used

$$f(F^{-1}(t)) = \lambda e^{(-\lambda)*(-\ln(1-t)/\lambda)} = \lambda e^{\ln(1-t)} = \lambda(1 - t).$$

Write t in the numerator as $1 - (1-t)$ and split the integral into two. They evaluate to $(1/e-1)/\lambda$ and $1/\lambda$. Add them together and apply the multiplier 2 to get the mean deviation as $2/(e\lambda)$. Alternatively put $y = 1-t$ in (2.12) and change the limits of integration as $1/e$ to 1 to get

$$MD = (2/\lambda) \int_{t=1/e}^{1} (1-y)/y \, dy \tag{2.13}$$

which readily evaluates to $2/(e\lambda)$.

MD of Laplace distribution

If X is Laplace(a, b) distributed, prove that the MD is b.

Solution The CDF of Laplace(a, b) is $y = F(x) = \frac{1}{2}\exp((x - a)/b)$, so that $x = F^{-1}(y) = a + b \ln(2y)$. From this we get $f(F^{-1}(x)) = x/b$. As $\mu = a$ (it's symmetric around a), $F(\mu) = 1/2$. Substitute in (2.11) to get $MD = 2\int_0^{1/2} t\,dt/(t/b) = 2b\int_0^{1/2}dt = b$.

MD of Uniform Distribution

If X is CUNI(a, b), find the mean deviation.

Solution As the CDF of CUNI(a, b) is $F(x) = (x-a)/(b-a)$, $F(\mu) = F((a + b)/2) = 1/2$ (this also follows trivially from the fact that CUNI(a, b) has special symmetry so that the area up to the mean is 1/2). As the density is constant throughout the range, $f(F^{-1}(t)) = 1/(b-a)$ always. Substitute these values in (2.11) to get

$$MD = 2 \int_{t=0}^{1/2} t \, dt/(1/(b-a)) = 2(b-a) \; t^2/2|_0^{1/2} = (b-a)/4. \qquad (2.14)$$

MD of Cosine distribution

If X is Cosine(a, b) distributed, prove that the MD is $b(\pi-2)$.

Solution Consider the Cosine(a, b) distribution with PDF

$$f(x; a, b) = (1/(2b)) \cos((x-a)/b) \qquad (2.15)$$

for $a - b\pi/2 \le x \le a + b\pi/2$. The CDF is

$$y = F(x) = (1/2)[\sin((x-a)/b) + 1], \qquad (2.16)$$

so that $x = F^{-1}(y) = a + b \; \sin^{-1}(2y-1)$. From this we get $f(F^{-1}(x)) = \sqrt{x(1-x)}/(2b)$. As $\mu = a$, $F(\mu) = F(a) = 1/2$ as the cosine distribution is symmetric [(Chattamvelli and Shanmugam 2020)]. Substitute in (2.11) to get $MD = 2\int_0^{1/2} t \, dt/f(F^{-1}(t)) = 2\int_0^{1/2} t dt/(\sqrt{t(1-t)}/(2b)) = 4b\int_0^{1/2} \sqrt{t/(1-t)}dt$. Put $t = \sin^2(\theta)$ so that $dt = 2 \sin(\theta)\cos(\theta) \, d\theta$. When $t = 1/2$, $\sin(\theta) = 1/\sqrt{2}$ so that $\theta = \pi/4$. The integrand becomes $\sin(\theta)/\cos(\theta)$. Cancel out $\cos(\theta)$ to get $8b\int_0^{\pi/4} \sin^2(\theta)d\theta$. Use $\sin^2(\theta) = (1 - \cos(2\theta))/2$ to get $MD = 4b\int_0^{\pi/4}(1 - \cos(2\theta))d\theta = 4b(\pi/4-1/2) = b(\pi-2)$.

2.3   Change of Variable Technique

The Change of Variable Technique (CoV-T) (also called Transformation of Variable Technique (ToV-T)) is a useful method to find distributions of simple continuous differentiable functions of real-valued RVs. It works on the principle that the average value of an integral $f_{avg}(x) = 1/(b-a) \int_a^b dF(x)$ can be equalized under an arbitrary continuous transformation $y = h(x)$ by the integral $g_{avg}(y) = 1/(h(b) - h(a)) \int_{h(a)}^{h(b)} dG(y)$ provided that $h(b) \to h(a)$ as $b \to a$. Equating the above gives

$$\int_a^b dF(x) = (b-a)/[h(b)-h(a)] \int_{h(a)}^{h(b)} dG(y). \qquad (2.17)$$

Now consider $\lim_{b \to a}[(h(b)-h(a))/(b-a)]$, which is the limiting value of the derivative of $h(x)$ at $x = a$. If this derivative $h'(x)$ exists for each point x in the interval (a, b), then the

RHS of (2.17) will be finite. When $b \to a$ from above, the LHS integral approaches f(x) and RHS integral approaches g(y)/h'(y). This allows us to equate the width of an infinitesimal strip under the function f(x) as f(x)dx = g(y)dy for all points x in (a, b). Because y = h(x) is invertible, we get g(y) = f(x)|$\partial h(y)/\partial y$| = |$1/J$|f(h^{-1}(y)) where J is called the Jacobian of the transformation. The only conditions in this transformation are that the mapping is once differentiable (i.e. h'(x) exists), and it is invertible (x can be expressed in terms of y). This can also be proved using the CDF method as follows.

Theorem 2.2 *The CDF of one-variable transformation is given by $G(y) = F(u^{-1}(y))$ if u(x) is strictly increasing and $1-F(u^{-1}(y))$ if u(x) is strictly decreasing. The PDF in both cases is $g(y) = f(u^{-1}(y))|\frac{\partial x}{\partial y}|$.*

Proof If u(x) is a strictly increasing function, and invertible,

$$G(y) = P[Y \le y] = P[u(x) \le y] = P[x \le u^{-1}(y)] = F(u^{-1}(y)). \qquad (2.18)$$

As the derivative is positive, the transformation results in g(y)∂y = f(x)∂x, so that g(y) = f(u^{-1}(y))|$\frac{\partial x}{\partial y}$|. If u(x) is strictly decreasing function, the derivative is negative so that the transformation results in G(y) = 1–F(u^{-1}(y)), and g(y) = –f(u^{-1}(y))|$\frac{\partial x}{\partial y}$|. This is the reason why the absolute value of the Jacobian (discussed in Chap. 3, Sect. 3.2) is taken. These results can easily be generalized to n-dimensions, as shown in the next chapter.

The standardization transformation Z = (X–μ)/σ where μ is the mean and σ is the standard deviation is the simplest and most frequent CoVT. The Jacobian in this case is |$\partial z/\partial x$| = 1/σ. When applied to an arbitrary normal distribution, this results in a standard normal distribution, which is extensively tabulated.

2.3.1   Linear Transformations

Any RV X can be transformed linearly using y = cx + b, where c $\ne$ 0. As they are linearly related, we could directly invert it to get x = (y–b)/c, and $|J| = 1/|c|$. Let $g_Y(y)$ denote the PDF of Y, and $f_X(x)$ denote the PDF of X. Then

$$g(y) = f(x)|\partial x/\partial y| = |J|f(h^{-1}(y)) = (1/|c|)f((y\text{–}b)/c), \qquad (2.19)$$

which is the same result obtained by the MoDF technique. When the variate is discrete, we simply ignore the 1/|c| multiplier. In general, if the transformation is Y = h(X), the PDF of Y is g(y) = f(h^{-1}(y)) if X is discrete and g(y) = f(h^{-1}(y))|$\partial h^{-1}(y)/\partial y$| if X is continuous.

Linear functions of binomial distribution

If X $\sim$ BINO(n, p) find distributions of Y = n–X, and find E(Y), V(Y).

Solution As Y = n–X takes integer values in reverse, it has the same distribution (see example in Chap. 1, Sect. 1.4.1). The PMF is f(y) = $\binom{n}{(n-y)}p^{n-y}q^{y}$. Using $\binom{n}{n-y} = \binom{n}{y}$, this can also be written as f(y) = $\binom{n}{y}q^{y}p^{n-y}$, y = 0, 1, . . . ,n; which is the PDF of a BINO(n, q). Hence E(Y) = nq, V(Y) = npq.

Linear functions of integral of PDF

If X is distributed as f(x; n) = $n(1-x)^{n-1}$, $0 < x < 1$, find the distributions of Y = $c(1-x)^{n}$ for c>0.

Solution It is easy to note that we are seeking the distribution of a constant times the integral of the PDF. Differentiation gives (dy/dx) – c $n(1-x)^{n-1}$. Use f(y) – f(x)/|(dy/dx)| and cancel out common terms to get g(y) = 1/c, 0 <y < c, which is CUNI(0, c).

2.4   Distribution of Sums

Sums of random variates appear in many engineering and applied science fields. They also arise in finding the distribution of likelihood ratio statistics. These are sometimes governed by physical laws like that between resistance and current in a circuit, and are usually represented as exact or approximate functional relationships among dependent and independent variables. If the distribution of one of them is known, the behavior of the other can easily be modeled. Most of the examples appearing in the following section assumes that the RVs are IID.

The distribution of sum of independent RVs is easy to find using convolution of integrals when both X and Y are continuous. Let $F_U(u)$ be the CDF of U. Then

$$F_U(u) = \Pr[U \leq u] = \int\int_{x+y\leq u} f(x)g(y)dxdy, \qquad (2.20)$$

due to the assumption of independence. As x + y≤ u represents the region below a straight line, the limits can be adjusted as $\int_{x=-\infty}^{\infty}\int_{y=-\infty}^{u-x} f(x)g(y)dxdy$. This can be factored as $\int_{x=-\infty}^{\infty}\left(\int_{y=-\infty}^{u-x} g(y)dy\right) f(x)dx$. Denote the inner integral by $G_Y(u - x)$. Above then becomes $\int_{x=-\infty}^{\infty} G_Y(u - x)f(x)dx$. Due to the symmetry of X and Y, we can rearrange the integrals to get a similar expression as $\int_{x=-\infty}^{\infty} F_X(u - y)g(y)dy$. Differentiation wrt u gives the PDF of X + Y as $g_U(u) =$

$$\frac{\partial}{\partial u}\int_{x=-\infty}^{\infty} G_Y(u - x)f(x)dx = \int_{-\infty}^{\infty} g(u - x)f(x)dx = \int_{-\infty}^{\infty} f(u - y)g(y)dy.$$
$$(2.21)$$

This is called the convolution of X and Y. The sum of uniform distributions with support (0, 1) follows directly using $f(u) = \int f_y(u - x) f_x(x) dx = \int_{max(0,u)}^{min(1,1+u)} dx = 1 + u$ for $-1 \leq u < 0$, and $1-u$ otherwise.

The distribution of difference can similarly be expressed as

$$Pr[X - Y \leq u] = Pr[X \leq Y + u] = \int_x Pr[X \leq y + u] f(y) dy = \int_x F_X(y + u) f_Y(y) dy.$$

(2.22)

Differentiate wrt u to get the PDF as $f_{X-Y}(u) = \int_x f_X(y + u) f_Y(y) dy$. The MGF is related as $M_{X+Y}(t) = M_X(t) * M_Y(t)$ and the CGF is related as $K_{X+Y}(t) = K_X(t) + K_Y(t)$ if X and Y are independent. This method can easily be extended to find the distribution of the difference of independent RVs.

Sum of Discrete Random Variables

If X is a discrete RV that takes values c and d (where c<d) with probabilities p and 1–p, and Y is an IID RV that takes values e and f (where e < f) with probabilities q and 1–q, find the PMF of U = X + Y, and its mean.

Solution The RV U = X + Y takes the values c + e, c + f, d + e and d + f with respective probabilities pq, p(1–q), (1–p)q and (1–p)(1–q) (see Fig. 2.1). Add them together to get [pq + p(1–q)] + [(1–p)q + (1–p)(1–q)] = p[q + 1–q] + (1–p)[q + 1–q] = p + 1–p = 1 showing that the probabilities add up to one. The mean is (c + e)pq + (c + f) p(1–q) + (d + e) (1–p)q + (d + f)(1–p)(1–q). This simplifies to cp + eq + d(1–p) + f(1–q) = d + f + p(c–d) + q(e–f).

2.4.1 Distribution of Sum of Non-identical RVs

The distribution of sum of non-identical RVs appear in several fields. These can easily be found when they are independent (in which case the joint PDF factors). As shown in the Jacobian section in Chap. 3, we could express the PDF of sum as

Fig. 2.1 Sum of discrete random variables

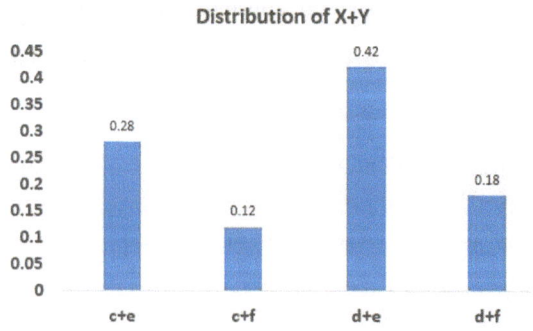

$$f_{X+Y}(u) = \int_{-\infty}^{\infty} f_{X,Y}(u-x)dx = \int_{-\infty}^{\infty} f(u-x)g(x)dx \qquad (2.23)$$

where the range of integration should subsume both the ranges (it must cover the minimum and maximum range of both).

Sum of Uniform and Standard Normal

If $X \sim$ CUNI$(-1, +1)$ and $Y \sim N(0, 1)$, find the PDF of $X + Y$.

Solution   Let $U = X + Y$. The PDF of X is $f(x) = 1/2$, $-1 \le x \le 1$ and that of Y is $f(y) = \frac{1}{\sqrt{2\pi}} \exp(-y^2/2)$. Use the above theorem to get the PDF as $f(u) = \int_{-\infty}^{\infty} f(u-x)g(x)dx$ $= \frac{1}{2\sqrt{2\pi}} \int_{-1}^{1} \exp(-(u-x)^2/2)dx$. As the densities are different from -1 to $+1$ and beyond, a point of inflection will occur at $x = -1$ and $x = +1$. This can be written in terms of Gaussian CDF as $\frac{1}{2}[\Phi(u+1) - \Phi(u-1)]$.

Sum of Poisson Variates

If X and Y are Poisson with parameters λ_1 and λ_2, find the PMF of $X + Y$.

Solution   We could easily get the required PMF using the MGF technique. But we proceed as follows to clarify the change of variable technique. The joint PMF of X and Y is $f(x, y) = e^{-\lambda_1}\lambda_1^x/x!e^{-\lambda_2}\lambda_2^y/y!$. Let $U = X + Y$, $V = Y$. The inverse mapping is $Y = V$, $X = U-V$. Thus $f_{U,V}(u, v) = e^{-\lambda_1}\lambda_1^{u-v}/(u-v)!e^{-\lambda_2}\lambda_2^v/v!$, where $v = 0, 1, 2, \ldots$ and $u = v, v+1, \ldots$ The PMF of U is obtained by summing over the entire range of V. As U has v as the lower bound, we need to sum over v from 0 to u. Hence $f_U(u) = e^{-(\lambda_1+\lambda_2)} \sum_{v=0}^{u} \lambda_1^{u-v}\lambda_2^v/[v!(u-v)!]$. Multiply and divide by $u!$ and take it outside the summation in the denominator. This gives $f_U(u) =$

$$\frac{e^{-(\lambda_1+\lambda_2)}}{u!} \sum_{v-0}^{u} \binom{u}{v} \lambda_1^{u-v}\lambda_2^v = \frac{e^{-(\lambda_1+\lambda_2)}}{u!}(\lambda_1 + \lambda_2)^u, u = 0, 1, 2 \ldots \qquad (2.24)$$

which is the PMF of POIS$(\lambda_1 + \lambda_2)$. Hence U is Poisson distributed.

Sum of Exponential Variates

If X and Y are independent exponential variates with the same shape parameter, find the distribution of $U = X + Y$.

Table 2.1 Joint distribution

X	Y	f(x, y)	Total
1	1	1/12	1/12
1	2	2/12	3/12
1	3	3/12	6/12
2	1	3/12	9/12
2	2	2/12	11/12
2	3	1/12	12/12

Table 2.2 Distribution of X + Y

X+Y	f(x, y)
2	1/12
3	5/12
4	5/12
5	1/12
Total	1.0

Solution We could use (2.21) to get the PDF. It is much easier to use the MGF. We know that the MGF of X is $(1 - t/\lambda)^{-1}$. As X and Y are independent, $M_{X+Y}(t) = M_X(t) * M_Y(t) = (1 - t/\lambda)^{-2}$, which is the MGF of gamma(2,λ).

PMF of sum X + Y

The joint PMF of X and Y is given in Table 2.1. Find the PMF of U = X + Y.

Solution Clearly, X + Y takes values in the range [2, 5].
P[X + Y = 2] = P[X = 1, Y = 1] = 1/12. P[X + Y = 3] = P[X = 1, Y = 2] + P[X = 2, Y = 1] = 2/12 + 3/12 = 5/12. Similarly, P[X + Y = 4] = P[X = 1, Y = 3] + P[X = 2, Y = 2] + P[X = 3, Y = 1] = 2/12 + 3/12 = 5/12, etc. The results are given in Table 2.2.

2.5   Distribution of Squares

Distribution of squares is especially important in statistical inference and analysis of variance. This is because we encounter sums of squares or functions thereof. For instance,
the ANOVA procedure is dependent on decomposing the total sum of squares as between
treatment and within sums of squares. Similarly, confidence intervals for variances are con-
structed using the distribution of sample variance, and testing of regression coefficients in
MLR uses the ratio of sums of squares. All of these require the distribution of appropriate
sums of squares under normality assumption. In such cases we need to find the distribution of
sums of independent normal random variates. Although the distribution of squares of other
RVs are seldom used in practice, they do have great theoretical significance. A special case
of the above is the relation between Student's T and Snedecor's F distributions. If $T \sim T_n$,
then the PDF of T_n^2 has an F distribution with 1 and n DoF. By definition, $T = Z/\sqrt{\chi_n^2/n}$ so
that $T_n^2 = Z^2/(\chi_n^2/n)$. As the numerator and denominator are independent, both of them are
chi-squared distributed so that their ratio has an F distribution.

There exist many methods to derive the distribution of squares. Let $Y = X^2$ so that dy/dx
= 2x and dx/dy = $1/(2\sqrt{y})$. If X takes positive values, the PDF of Y is

$$g(y) = f(x)|\partial x/\partial y| = f(\sqrt{y})/(2\sqrt{y}). \tag{2.25}$$

The corresponding relationship between distribution functions for strictly increasing functions is

$$G(y) = \Pr(X^2 \le y) = \Pr[-\sqrt{y} \le x \le \sqrt{y}] = F(\sqrt{y}) - F(-\sqrt{y}). \tag{2.26}$$

As (2.26) is valid for any x($-\infty < x < \infty$), we differentiate it wrt y to get the PDF as

$$g(y) = (f(\sqrt{y}) + f(-\sqrt{y}))/(2\sqrt{y}). \tag{2.27}$$

Distribution of the square of a T variate

If X is T(n), find the distribution of X^2/n.

Solution The PDF of Student's T distribution is

$$f(t) = K(1 + t^2/n)^{-(n+1)/2} \tag{2.28}$$

where K = $\Gamma((n+1)/2)/[\sqrt{n\pi}\ \Gamma(n/2)]$. Using (2.26), the PDF of Y = T^2/n is obtained as
G(y) = Pr($Y \le y$) =

$$\Pr(T^2/n \le y) = \Pr(-\sqrt{ny} \le T \le \sqrt{ny}) = F(\sqrt{ny}) - F(-\sqrt{ny}). \tag{2.29}$$

Differentiate (2.29) wrt y to get the PDF of Y as

$$\sqrt{n}[f(\sqrt{ny}) + f(-\sqrt{ny})](1/(2\sqrt{y})).$$ (2.30)

As the T distribution is symmetric, $f(\sqrt{y}) = f(-\sqrt{y})$. Substitute in (2.28), and cancel out $\sqrt{n}$ to get the desired PDF as

$$g(y, n) = \Gamma((n + 1)/2)/[\sqrt{\pi}\ \Gamma(n/2)](1 + y)^{-(n+1)/2}/\sqrt{y}.$$ (2.31)

Write $\sqrt{y}$ in the denominator as $y^{1/2-1}$ and take numerator expression to the denominator. Then this is found to be a BETA-II distribution

$$(1/B(1/2, n/2))y^{1/2-1}/(1 + y)^{(n+1)/2},$$ (2.32)

where $B(1/2, n/2) = \Gamma((n + 1)/2)/[\sqrt{\pi}\ \Gamma(n/2)]$ is the CBF.

Distribution of the square of Rayleigh variate

If X is Rayleigh(a) with PDF $f(x, a) = (x/a^2) \exp(-x^2/(2a^2))$, find the distribution of X^2.

Solution Let $y = x^2$ so that dy/dx = 2x, or $x\,dx = dy/2$. Then $f(y) = f(x)*dx/dy = (1/(2a^2))\exp(-y/(2a^2))$, which is an exponential distribution $EXP(1/(2a^2))$. Then

Problem 2.1 If $X \sim N(0, \sigma^2)$, prove that X^2 has a scaled χ_1^2 distribution.

2.6   Distribution of Sum of Squares

First consider the sum of squares of two IID variates. Let $Z = X^2 + Y^2$ where X and Y are independent. If X and Y are IID N(0, 1), their squares are χ_1^2 and the sum of squares is χ_2^2 (central chi-square with 2 DoF) distributed. Otherwise proceed as follows:

$$F(z) = \Pr(Z \leq z) = \Pr(X^2 + Y^2 \leq z) = \int\int_{X^2+Y^2 \leq z} f(x, y)dxdy.$$ (2.33)

As $X^2 + Y^2 \leq z$ represents the area enclosed by a circle of radius $\sqrt{z}$ centered at the origin, we could make a polar transformation $x = r\cos(\theta)$, $y = r\sin(\theta)$ (and $|J| = r$, see Table in next chapter) to get

$$F(r, \theta) = \int_{-\sqrt{z} < r < \sqrt{z}} \int_{-\pi/2 < \theta < \pi/2} f(r, \theta)r\ dr\ d\theta.$$ (2.34)

From this the distribution of r is found by integrating out θ and then the distribution of $r^2 = x^2 + y^2$ can easily be found. Alternatively, write (2.33) as

$$F(z) = \int_{y=-\sqrt{z}}^{\sqrt{z}} \left(\int_{-\sqrt{z-y^2}}^{\sqrt{z-y^2}} f(x, y)dx \right) dy.$$

(2.35)

Differentiate (2.35) wrt z and use Leibniz theorem on the RHS to get the PDF as

$$f(z) = \frac{1}{2} \int_{-\sqrt{z}}^{\sqrt{z}} (1/\sqrt{z-y^2}) \left(f(\sqrt{z-y^2}, y) + f(-\sqrt{z-y^2}, y) \right) dy.$$

(2.36)

As $\sum_{i=1}^{n} x_i^2 = r^2$ represents a hypersphere of radius r in n-dimensions, an exactly similar procedure can be used when the variates are independent. For dependent variates, we may use an orthogonal transformation to a new set of variables with known distribution and derive the distribution of original variates.

2.7 Distribution of Square-Roots

As the square root of a negative number is imaginary, this transformation is defined only for RVs that take positive values. It makes sense for continuous variates than discrete ones. Let $Y = \sqrt{X}$, which gives $X = Y^2$ and dx/dy = 2y. The straightforward way to find the PDF of Y is to use

$$g(y) = f(x)|\partial x/\partial y| = 2yf(y^2).$$

(2.37)

If the resulting distribution of Y is symmetric, we need to divide the final density by 2 to get the correct PDF, because both $(-y)^2$ and $(+y)^2$ map to x. Symbolically, g(y) = y f(y²) if Y is symmetric. This is summarized as

$$g(y) = f(x)|\partial x/\partial y| = \begin{cases} 2yf(y^2) & \text{if } Y \text{ is asymmetric;} \\ yf(y^2) & \text{if } Y \text{ is symmetric.} \end{cases}$$

Distribution of the square-root of a χ^2 variate

If X is χ_n^2, find the distribution of $\sqrt{X}$.

Solution The PDF of χ_n^2 variate is $f(x) = e^{-x/2} x^{n/2-1}/[2^{n/2}\Gamma(n/2)]$ where n is the DoF ≥ 1. Put $Y = \sqrt{X}$ and use (2.37) to get

$$f(y) = 2y * e^{-y^2/2}(y^2)^{n/2-1}/[2^{n/2}\Gamma(n/2)].$$

(2.38)

As $y*(y^2)^{n/2-1} = y^{n-1}$ the PDF becomes $f(y) = e^{-y^2/2}y^{n-1}/[2^{n/2-1}\Gamma(n/2)], 0 < y < \infty.$
This is the chi-distribution , or the standard form of Rayleigh distribution.

Distribution of the square-root of an F variate

If X is F(1, n), find the distribution of $\sqrt{X}$.

Solution The PDF of F distribution is

$$f(x; m, n) = \frac{\Gamma((m + n)/2)m^{m/2}n^{n/2}}{\Gamma(m/2)\Gamma(n/2)} \frac{x^{m/2-1}}{(n + mx)^{(m+n)/2}}, \quad 0 < x < \infty. \tag{2.39}$$

Put m = 1 to get the PDF of F(1, n) as

$$f(x; n) = \frac{\Gamma((1 + n)/2)n^{n/2}}{\Gamma(1/2)\Gamma(n/2)} \frac{x^{1/2-1}}{(n + x)^{(1+n)/2}}, \quad 0 < x < \infty. \tag{2.40}$$

Now use (2.37) to get g(y) =

$$2yf(y^2) = 2y\Gamma((1 + n)/2)n^{n/2}/[\Gamma(1/2)\Gamma(n/2)](y^2)^{1/2-1}/(n + y^2)^{(1+n)/2} \tag{2.41}$$

The y cancels out with $(y^2)^{1/2-1} = 1/y$. Take n outside from the bracket in the denominator
and cancel out with $n^{n/2}$ in the numerator to get a $\sqrt{n}$ in the denominator. As the PDF now
involves only powers of y^2, it is symmetric. Hence we need to divide the resulting PDF by
2 to get the correct PDF as

$$g(y) = (1 + y^2/n)^{-(n+1)/2}/[\sqrt{n}B(1/2,n/2)] \tag{2.42}$$

which is the Student's T(n) distribution.

Distribution of the square-root of U(0, 1) variate

If X is U(0, 1), find the distribution of $Y = \sqrt{X}$.

Solution We know that the square root of a number less than one is greater than itself. As
the square root of all fractions in [0, 1] belongs to the same interval, the domain of Y is also
[0, 1]. Using Eq. (2.42), the PDF of Y is 2 y f(y^2) = 2y for 0<y<1. This is twice the CDF
of U(0, 1). For y > 0.5, the PDF of Y will attain values greater than one (see discussion in
Chap. 1, Sect. 1.2.2).

2.8   Distribution of Reciprocals

Distribution of reciprocals is defined only in some particular cases. If the value of a RV X at x = 0 is nonzero, the RV Y = 1/X is well-defined. It can also be used (along with Sect. 2.3.1) to find the distribution of (X–1)/X = 1–1/X and (1–X)/X = 1/X–1. The straightforward way to find the PDF of Y is to use

$$g(y) = f(x)|\partial x/\partial y| = f(1/y)/y^2. \tag{2.43}$$

Distribution of the reciprocal of a Cauchy variate

If X is Cauchy distributed, find the distribution of Y = 1/X.

Solution The PDF of Cauchy distribution is $f(x) = \frac{1}{\pi}\frac{1}{1+x^2}$, $-\infty < x < \infty$. As f(x = 0) is $1/\pi$, distribution of the reciprocal is well-defined. Using (2.43) the PDF becomes $f(y) = \frac{1}{\pi}\frac{1}{1+(1/y)^2}/y^2$. The y^2 cancels out from the numerator and denominator, giving $f(y) = \frac{1}{\pi}\frac{1}{1+y^2}$, which is Cauchy distributed.

Reciprocal of a unit rectangular variate

If X is U(0, 1) distributed, find the distribution of Y = 1/X.

Solution Let G(y) be the CDF of Y. Then

$$G(y) = \Pr[Y \le y] = \Pr[1/X \le y] = \Pr[X \ge 1/y] = 1\text{-}1/y. \tag{2.44}$$

Differentiate wrt y to get the PDF as g(y) = $1/y^2$, for y≥1.

2.9   Distribution of Minimum and Maximum

The distribution of minimum and maximum (called extremes) finds applications in many fields. For example, distribution of maximum is used in life sciences to model the survival time of species, produce, machines, and various products. It is also used in reliability theory to model the life of equipments and parts, various devices, and consumer items (like light bulbs and computer chips). The study of extremes is called order statistics. Let $X_1, X_2, \ldots, X_n$ be a random sample from an arbitrary distribution with PDF f(y) and CDF F(y). Let $Y_1 = \min(X_1, X_2, \ldots, X_n)$. Then

$$1 - F_{Y_1}(y) = \Pr(Y_1 > y) = \Pr(X_1 > y) * \cdots * \Pr(X_n > y) = [1 - F(y)]^n \tag{2.45}$$

using independence. Differentiate the above to get the PDF of Y_1 as

$$f(y_1) = n(1 - F(y))^{n-1} f(y). \tag{2.46}$$

Similarly, the PDF of y_n using CDF method is $f(y_n) = n\, f(y)\, [F(y)]^{n-1}$.

Distribution of minimum of GEO(p)

Prove that the distribution of the minimum of n IID geometric RVs has a geometric distribution.

Solution Consider $\Pr[Z>z] = \Pr[\min(X,Y)> z] = \Pr[X \geq z, Y \geq z] = \Pr[X \geq z]\,\Pr[Y \geq z]$ (due to independence). As the SF of a geometric distribution GEO(p) is $(1 - p)^x$ [(Chattamvelli and Shanmugam 2020)] the above becomes $(1 - p)^z (1 - q)^z = [(1 - p)(1 - q)]^z$. Hence $\Pr[Z \leq z] = 1 - [(1 - p)(1 - q)]^z$, which is the CDF of a geometric distribution with parameter $1 - (1-p)(1-q) = p + q - pq$. Hence the resulting distribution is geometric. This result can easily be extended to n RVs. First consider the case where each of them is Geo(p) (with same p). Let $Z = \min(X_1, X_2, \ldots, X_n)$. Then $\Pr[Z>z] = \Pr[\min(X_1, X_2, \ldots, X_n) > z]$. This is the same as the probability that each of them is $>z$, which follows by independence assumption as $(1 - p)^{nz}$. This is the SF of a GEO$(1-(1-p)^n)$. Next consider the case where each of them is GEO(p_k). An exactly similar steps will show that $Z \sim$ GEO$(1-(1 - p_1)(1 - p_2) \cdots (1 - p_n))$.

2.10   Distribution of Trigonometric Functions

Trigonometric functions of some RVs are easy to work with. One example is the Cauchy distribution. If X has a standard Cauchy distribution, then Cos(X) has the same distribution. Trigonometric functions are also utilized to derive some distributions using geometric concepts. One example is the correlation coefficient. The cosine of the angle between two normalized vectors in n-dimensional Euclidean space is called the correlation coefficient.

Distribution of U = tan(X)

If X has a U(0, 1) distribution, find the distribution of U = tan(X).

Solution The PDF of X is $f(x) = 1, 0< x <1$. The inverse transformation is $X = \tan^{-1}(U)$. This gives $|\partial x / \partial u| = 1/(1 + u^2)$. The range of U is modified as $\tan(0) = 0$ to $\tan(1) = \pi/4$. Hence the distribution of U is $f(u) = 1/(1 + u^2)$ for $0< u < \pi/4$.

tan(1/X) of uniform distribution

If X is CUNI[0, 1], find the distribution of U = c tan(1/X).

Solution U = c tan(1/X) gives X = $1/\tan^{-1}(U/c)$ and dx/du = $-1/c[\tan^{-1}(U/c)]^2$. As X is CUNI[0, 1] (U[0, 1], unit normal) f(x) = 1 so that f(u) = $(1/c)[\tan^{-1}(u/c)]^2$ for u ≥ c.

Distribution of U = sin(X)

If X has a CUNI[$-\pi/2$, $\pi/2$] distribution, find the distribution of U = sin(X).

Solution The inverse transformation is x = $\sin^{-1}(u)$ so that $|\partial x/\partial u| = 1/\sqrt{1-u^2}$. When x = $-\pi/2$, u = $\sin(-\pi/2) = -\sin(\pi/2) = -1$. When x = $\pi/2$, u = $\sin(\pi/2) = 1$. As the PDF of X is $1/\pi$, the PDF of U is f(u) = $(1/[\pi\sqrt{1-u^2}]$, for $-1 < u < +1$.

Distribution of Perpendicular height

Let A be an arbitrary point on the circumference of a uniformly rotating wheel with center O and radius r. Find the distribution of the height h from A to the horizontal axis.

Solution Let θ be the angle made by the radius OA to the positive horizontal axis, and P be the touching point on the axis so that POA is a right triangle with hypotenuse OA (Fig. 2.2). Assume without loss of generality that the point A is in the first quadrant. As the wheel makes one complete revolution, θ will vary from 0 to 2π. Due to the symmetry of the circle, there exist four positions in which the vertical distance from a point on the circumference to the horizontal axis are equal (reflection of A in 4th quadrant, at position B = $(\theta + \pi/2)$ in the second quadrant, and its reflection point in third quadrant). If we restrict θ to vary from $-\pi/2$ to $\pi/2$ (which covers first and fourth quadrants) we need consider two such points. From the triangle, we have h = r sin(θ) which varies monotonically. Now proceed exactly as in the above example using h = r sin(θ) to get $|\partial\theta/\partial h| = 1/[r\sqrt{1-(h/r)^2}] = r/\sqrt{r^2-h^2}$. The PDF of h becomes (Fig. 2.3)

$$g(h, r) = r/[\pi\sqrt{(r^2-h^2)}], \text{ for } -r < h < r.$$ (2.47)

Due to the symmetry mentioned above, the distribution of t = $|h|$ becomes

$$g(t, r) = 2r/[\pi\sqrt{r^2-t^2}], \text{ for } 0 < t < r.$$ (2.48)

Fig. 2.2 Distribution of
perpendicular height PA

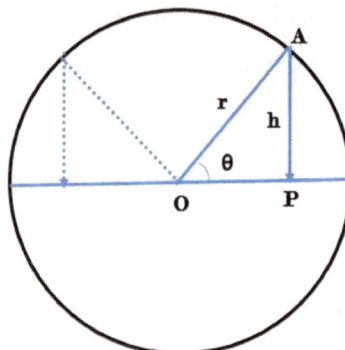

Fig. 2.3 Distribution of
perpendicular height

Distribution of $g(h, r) = r/(\pi\sqrt{r^2 - h^2})$ for r=3, $-r < h < +r$

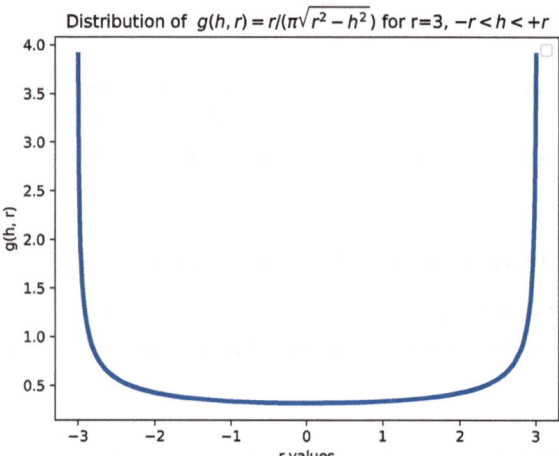

2.11   Distribution of Transcendental Functions

Distributions of transcendental functions are quite useful in engineering and related fields. Several laws and principles in engineering and physical sciences are modeled as mathematical equations called functionals involving the unknown variables and known constants. In most applications, one variable (called dependent variable) is modeled as a function of two or more other variables (called independent variables), which can include time. If the number of variables involved is two, the CoVT technique is useful to derive the distribution of one, using the distribution of the other.

2.11.1 Distribution of Logarithms

The logarithmic transformation can be applied to any RV that takes non-negative values. As $\log(0) = -\infty$, this transforms $(0, \infty)$ to the new range $(-\infty, \infty)$. As the log() is a real function of its argument, this transformation is applied to continuous RVs. Unless otherwise stated, the base of the logarithm is assumed to be e. A special transformation encountered in communication theory is $Y = \log_2(1 + X)$. Using the method described above, the PDF of Y is given by $g_Y(y) = \ln(2) f(2^y - 1)2^y$ (Table 2.3).

Logarithmic transformation of CUNI Distribution

If X is CUNI[a, b] distributed, find the distribution of Y = –log(X–a)/(b–a).

Solution The CDF of Y is $F(y) = \Pr[Y \leq y] = \Pr[(X–a)/(b–a) \geq e^{-y}] = \Pr[X \geq a + (b-a)e^{-y}]$. As the CDF of CUNI(a, b) is (x–a)/(b–a) this becomes $1 - \Pr[X \leq a + (b-a)e^{-y}] = 1-e^{-y}$. From this the PDF is obtained by differentiation as $f(y) = e^{-y}$. Hence Y is EXP(1).

Logarithm of Weibull distribution

If $X \sim$ Weibull(a, b), find the distribution of Y = log(X/b).

Solution The PDF of two-parameter Weibull distribution is $f(x, a, b) = \frac{a}{b}(x/b)^{a-1}e^{-(x/b)^a}$ where b is the scale and a is the shape parameter. As $Y = \log(X/b)$ we have dy/dx = 1/x, and x = b exp(y). From this we get $f(y) = \frac{a}{b}^2 \exp(y)(\exp(y))^{a-1}e^{-(\exp(y))^a}$.

Exponentiated exponential

If X is EXP(λ), find the distribution of $Y = e^X$.

Solution The PDF of X is $f(x) = \lambda \exp(-\lambda x)$ for $x \geq 0$. This has CDF $1-\exp(-\lambda x)$ [Chattamvelli and Shanmugam (2021)]. Let F(y) denote the CDF of $Y = e^X$. The lower limit of Y is obviously $e^0 = 1$ and upper limit is ∞. By definition $F(y) = \Pr(Y \leq y) = \Pr(e^X \leq y)$. As exp(x) is an increasing function, take log (to base e; or ln) of both sides to get $F(y) = \Pr(X \leq \ln(y)) = F_X(\ln(y)) = 1-\exp(-\lambda \ln(y))$. Now use $\log(a^b) = b \log(a)$, and exp(ln(x))=x to get $F(y) = 1- y^{-\lambda}$. Differentiate wrt y to get the PDF as $f(y) = \lambda y^{-\lambda-1} = \lambda/y^{1+\lambda}, 1 \leq y < \infty$.

Table 2.3 Summary table of transformation of variates

Transformation y = h(x)	Transformed density function	Comments/Conditions
$c*x + d$	$\frac{1}{|c|} f((y-d)/c)$	$c \neq 0$, not near 0
$|x|$	$g(y) = f(y) + f(-y)$	$g(y) = 2f(y)$ if Y symmetric
$|x-c|$	$g(y) = f(c+y) + f(c-y)$	$g(y) = 2f(c+y)$ if Y is symm.
x^2	$f(\sqrt{y})/(2\sqrt{y})$	x positive
x^2	$[f(\sqrt{y}) + f(-\sqrt{y})]/(2\sqrt{y})$	$-\infty < x < \infty$
cx^2	$[f(\sqrt{y/c}) + f(-\sqrt{y/c})]/(2\sqrt{y/c})$	Any x
$1/[cx^2]$	$1/[2\sqrt{c}y^{3/2}]f(1/\sqrt{cy})$	
$\sqrt{x}$	$2y\, f(y^2)$	$yf(y^2)$ if Y is symmetric
$\sqrt{cx}$	$2(y/c)f(y^2/c)$	$2y\, f(y^2)$ if c = 1
$1/x$	$f(1/y)/y^2$	For $x \neq 0$
x^α	$\frac{1}{\alpha}f(y^{1/\alpha})y^{1/\alpha-1}$	$0 < \alpha < 1, x > 0$
x^n	$\frac{y^{1/n-1}}{n}\left[f(y^{1/n}) + f(-y^{1/n})\right]$	For y > 0
e^x	$f(\ln(y))/y$	Log to base e
e^{ax}	$f(\ln(y)/a)/(ay)$	Log to base e
$\tan^{-1}(x)$	$f(\tan y)\sec^2(y)$	x in radians
$-2\ln(x)$	$\frac{1}{2}e^{-y/2}f(e^{-y/2})$	Log base e, $x \geq 0$
$-\ln(1-x)/\lambda$	$\lambda e^{-\lambda y}f(1 - e^{-\lambda y})$	Log base e, $0 < x < 1$
$1/[1 + e^{-x}]$	$f(\ln(y/(1-y)))/[y(1-y)]$	Std. sigmoid function
$x/(1-x)$	$f(y/(1+y))/(1+y)^2$	$0 < x < 1$
$\ln(x/(1-x))$	$f(e^y/(e^y+1))e^y/(1+e^y)^2$	$0 < x < 1$
$\ln[(1+x)/(1-x)]$	$f((e^y - 1)/(e^y + 1))2e^y/(1+e^y)^2$	$0 < x < 1$
$\log_2(1+x)$	$ln(2)\,2^y f(2^y - 1)$	x is non-negative

The ranges are not shown above as it depends on the range of the original variate. Note that log x and $\sqrt{x}$ are concave functions

Transformation of Arc-Sine Distribution

If X is Arc-Sine distributed, find the distribution of Y = −log(X).

Solution Let F(y) be the CDF of Y. Then F(y) = Pr[Y≤ y] = Pr[X≥ e^{-y}] = 1−(2/π) $\sin^{-1}(\sqrt{e^{-y}})$. Differentiate wrt y to get the PDF as f(y) = (2/π) $e^{-y/2}/[2(\sqrt{1 - e^{-y}})]$. The 2 cancels out from numerator and denominator giving

$$f(y) = (1/\pi)e^{-y/2}/\sqrt{1 - e^{-y}}, \quad 0 \leq y < \infty. \tag{2.49}$$

Transformation of Cauchy Distribution

If X is Cauchy distributed, find the distribution of $Y = \tan^{-1}(X)$.

Solution Differentiate wrt x to get $dy/dx = 1/(1 + x^2)$. Hence $f(y) = f(x) |dx/dy| = \frac{1}{\pi}$. As $\tan(\pi/2) = \infty$, the range is $(-\pi/2, \pi/2)$.

2.11.2 Distribution of Arbitrary Functions

Consider the transformation $y = g(x)$ where $g(x)$ is a one-one mapping that is invertible. This means that x can be expressed in terms of Y as $x = g^{-1}(y)$. Then the PDF of Y can be represented in the continuous case as follows. First express the CDF of Y in terms of the CDF of X as

$$F_Y(y) = \Pr(Y \leq y) = \Pr(g(x) \leq y) = \Pr(x \leq g^{-1}(y)) = F_X(g^{-1}(y)). \qquad (2.50)$$

Differentiate both sides to get the PDF as

$$f_Y(y) = f_X(g^{-1}(y))\partial g^{-1}(y)/\partial y \qquad (2.51)$$

and in the discrete case $Y = r(X)$ as

$$f_Y(y) = \Pr_X(r(x) = y) = \sum_{x:r(x)=y} \Pr_X(X = x). \qquad (2.52)$$

Distribution of $Y = -\ln(X/(1-X))$

If X is BETA–I(p, q) distributed, show that the variate $Y = -\ln(X/(1-X))$ is type-2 generalized logistic(p, q) distributed.

Solution Let $X/(1-X) = \exp(-Y)$ so that $x = \exp(-Y)/(1 + \exp(-Y))$ and $1-X = 1/(1 + \exp(-Y))$. This gives $dy/dx = 1/[x(1-x)]$. Substitute the values in the PDF of BETA–I and multiply by absolute value of Jacobian to get $f(y) = (1/B(a, b))[\exp(-Y)/(1 + \exp(-Y))]^{a-1}[1/(1 + \exp(-Y))]^{b-1} * \exp(-Y)/(1 + \exp(-Y))^2$. This simplifies to $\exp(-ay)/(1 + \exp(-y))^{a+b}$ which is type-2 generalized logistic law (Table 2.3).

X^b of Weibull distribution

If X has standard Weibull distribution with PDF $f(x) = bx^{b-1}e^{-x^b}$, find the distribution of $Y = X^b$.

Solution As $Y = X^b$, $dy/dx = bx^{b-1}$ and $x = y^{1/b}$. From this $f(y) = \exp(-y)$, which is the standard exponential distribution.

Distribution of Y = 1/(1 + x)^c.

If $f(x) = (c-1)/(1 + x)^c$ for $0 < x < \infty$, find the distribution of $Y = 1/(1 + x)^c$ and find the mean and variance.

Solution As $Y = 1/(1+x)^c$, $dy/dx = -c/(1 + x)^{c+1} = -cy^{1+1/c}$. From this $f(y) = (c-1)/(1 + x)^c * (1 + x)^{c+1}/c = (c - 1)/cy^{-1/c}$, $0 < y < 1$. The mean is $E(Y) = (c-1)/c \int_0^1 y^{1-1/c} dy = (c-1)/c*c/(2c-1) = (c-1)/(2c-1)$. $E(Y^2) = (c-1)/c \int_0^1 y^{2-1/c} dy = (c-1)/c*c/(3c-1) = (c-1)/(3c-1)$. From this variance is obtained as $(c-1)/(3c-1)-(c - 1)^2/(2c - 1)^2$.

2.12   Distribution of Integer and Fractional Parts

Integer and fractional parts are applicable to continuous RVs, usually univariate RVs with infinite range. Integer part gives rise to discrete distribution whereas fractional part results in continuous distribution. It is applicable to classical, truncated, size-biased and other types of distributions.

2.12.1 Distribution of Integer Part

If X is Cauchy distributed, it is easy to find the distribution of the integer part $Y = \lfloor X \rfloor$. We have $f(x) = 1/[\pi(1 + x^2)]$ for $-\infty < x < \infty$. The RV Y takes integer values on the entire real line including $y = \mp\infty$. Specifically,

$$\Pr[Y = y] = \Pr[y \leq X < y + 1] = \int_y^{y+1} 1/[\pi(1 + x^2)]dx = (1/\pi) \tan^{-1}(x)|_y^{y+1}$$

$$= (1/\pi)[\tan^{-1}(y + 1) - \tan^{-1}(y)].$$

A similar expression could be obtained for $x < 0$ as $\Pr[Y = y] = \Pr[y-1 \leq X < y]$. As alternate terms of the above equation cancel out, this form is useful to compute the CDF, and to prove that the probabilities add up to 1. For example, sum from $-\infty$ to ∞ to get $\tan(\infty)$-tan$(-\infty) = \pi/2 - (-\pi/2) = \pi$, which cancels with π in the denominator. Now use the identity $\tan^{-1}(x) - \tan^{-1}(y) = \tan^{-1}((x - y)/(1 + xy))$ to get

$$\Pr[Y = y] = (1/\pi) \tan^{-1}(y + 1 - y)/[1 + y(y + 1)] = (1/\pi) \tan^{-1}(1/[1 + y(y + 1)]), \tag{2.53}$$

which is the desired form. As the terms do not cancel, this form is not useful to compute the CDF.

This example shows another way to specify a PDF as the difference of two functions (say $[\tan^{-1}(x + 1) - \tan^{-1}(x)]$ or $[\exp(-\lambda x) - \exp(-\lambda(x + 1))]$) that simply cancels out when summed over the proper range of x, leaving behind only two extreme terms whose difference appears as the normalizing constant in the denominator. This allows us to define a variety of new discrete distributions. As the PMF should be positive, we should form the difference separately for positive and negative values, to make it non-negative. The shape of the distribution (whether it is unimodal and tails off to the extremes, or it is U-shaped or J-shaped etc.) must be known to form the PDF.

2.12.2 Distribution of Fractional Part

If X has an Exponential distribution, the distribution of $Y = \lfloor X \rfloor$ is GEO$(1 - \exp(-\lambda))$ [Chattamvelli and Shanmugam (2021)]. The values of $Y = X - \lfloor X \rfloor$ are in $0 \leq y \leq 1$. If we assume that the integer and fractional parts are independent, Y is the difference between an exponential and geometric RVs. This is of mixed type (as geometric distribution is discrete), where the continuous distribution dominates. This means that Y has a continuous distribution. Using the MoDF it is easy to show that Y is distributed as f(y) = $\lambda \exp(-\lambda y)/[1 - \exp(-\lambda)]$, for $0 \leq y \leq 1$.

2.12.3 Special Functions

There exist many symmetric and skew symmetric functions that possess interesting properties. These can be in single or multiple variables. Examples are $x/(1 - x)$, $(1 + x)/(1 - x)$, $e^{-x}/(1 + e^{-x})$, etc. Consider $Y = X/\sqrt{1 + X^2}$. Square both sides to get $Y^2 = X^2/(1 + X^2)$, from which $x = y/\sqrt{1 - y^2}$. Differentiate wrt y to get $|J| = |\partial x/\partial y| = 1/(1 - y^2)^{3/2}$. From this the PDF of Y is easy to obtain as

$$g(y) = f(y/\sqrt{1 - y^2})/(1 - y^2)^{3/2}. \tag{2.54}$$

If X is standard Cauchy distributed, then $Y = X/\sqrt{1 + X^2}$ has PDF

$$g(y) = (1/\pi)(1 - y^2)/(1 - y^2)^{3/2} = (1/[\pi(1 - y^2)^{1/2}], \quad -1 \leq y \leq +1. \tag{2.55}$$

2.12.4 Distribution of Ratio of Sums

Probability distribution of ratio of sums and ratio of sums-of-squares appear at various places such as ANOVA, ANCOVA and MLR. Let X_i's be IID EXP($\lambda\theta$) for i = 1, 2, ..., m. Let Y_j's be IID EXP(θ) for j = 1, 2, ..., n. If X_i's and Y_j's are pair-wise independent, we could find the distribution of the ratio W = U/V = $\sum_{i=1}^{m} X_i / \sum_{j=1}^{n} Y_j$ as follows. As X_i's are IID, the joint PDF is the product of individual PDFs. The MGF technique can be used to find the distribution of numerator and denominator. The MGF of EXP($\lambda\theta$) is $M_x(t) = 1/[1 - \lambda\theta t]$. As the X_i's are IID, $M_u(t) = 1/[1 - \lambda\theta t]^m$. Similarly, $M_v(t) = 1/[1 - \theta t]^n$. These are the MGFs of gamma distributions. Hence W is the ratio of two independent gamma variates.

2.13   Transformations of Normal Variates

Most of the transformations discussed above are applicable to the normal variate. As this has many practical applications, this section briefly discusses some of them.

2.13.1 Linear Combination of Normal Variates

Linear combination of any number of independent normal variates is normally distributed. This can be proved using induction. A simpler method is to use the MGF technique. The ChF of N(μ, σ^2) is $\exp(it\mu - \frac{1}{2}t^2\sigma^2)$. If $X_1, X_2, \ldots, X_n$ are independent normal RVs N(μ_i, σ_i^2), the ChF of Y = $X_1 + X_2 + \cdots + X_n$ is

$$\phi_y(t) = \prod_{i=1}^{n} \phi_{x_i}(t) = \exp\left(it \sum_i \mu_i - \frac{1}{2}t^2 \sum_i \sigma_i^2\right). \tag{2.56}$$

As this is the ChF of a normal variate with mean $\sum_i \mu_i$ and variance $\sum_i \sigma_i^2$, it follows that Y is normally distributed. In the particular case when each of the μ_i's and σ_i^2 are equal, we have N($n\mu$, $n\sigma^2$). If Y = $c_1 X_1 + c_2 X_2 + \cdots + c_n X_n$, the ChF of Y is

$$\phi_y(t) = \prod_{i=1}^{n} \phi_{c_i x_i}(t) = \prod_{i=1}^{n} \phi_{x_i}(c_i t) = \exp\left(it \sum_i c_i \mu_i - \frac{1}{2}t^2 \sum_i c_i^2 \sigma_i^2\right). \tag{2.57}$$

This shows that $\sum_i c_i Y_i$ is normal with mean $\mu = \sum_i c_i \mu_i$, and variance $\sigma^2 = \sum_i c_i^2 \sigma_i^2$. It follows that cX+b is normal with mean $\mu' = c * \mu + b$ and variance $\sigma'^2 = c^2 \sigma^2$.

2.13.2 Square of Normal Variates

The square of a normal variate is χ_1^2 distributed. In general the sum of squares of any number of independent normal variates is χ_n^2 distributed. Similarly, the ratio of independent normal variates is Cauchy distributed. This is proved as follows.

Distribution of the square of a normal variate

If X is N(0, 1), find the distribution of X^2.

Solution Using the MoDF we get G(y) =

$$\Pr(Y \le y) = \Pr(X^2 \le y) = \Pr(-\sqrt{y} \le X \le \sqrt{y}) = \int_{-\sqrt{y}}^{\sqrt{y}} (1/\sqrt{2\pi})e^{-z^2/2}dz. \quad (2.58)$$

Differentiate wrt y to get

$$g(y) = (1/\sqrt{2\pi}) \frac{\partial}{\partial y} \int_{-\sqrt{y}}^{\sqrt{y}} e^{-z^2/2}dz. \quad (2.59)$$

Using Leibniz theorem this reduces to $(1/\sqrt{2\pi}) \left(e^{-y/2} * 1/(2\sqrt{y}) - e^{-y/2} * (-1)/(2\sqrt{y})\right)$ $= (1/\sqrt{2\pi})y^{1/2-1}e^{-y/2}$. This is the PDF of a chi-square variate with 1 DoF. Alternatively, write (2.59) as $\frac{\partial}{\partial y}[\Phi(\sqrt{y}) - \Phi(-\sqrt{y})]$ and proceed as above.

2.13.3 Other Transformations of Normal Variates

The ratio of IID normal variates is Cauchy distributed (Chap. 3, Sect. 3.3.3). The ratio of a normal over an independent uniform distribution is called *slash distribution*. If X is normally distributed, then Y = exp(X) has a log-normal distribution. If X,Y are IID N(0, 1) distributed, then $2XY/\sqrt{X^2+Y^2}$ and $(X^2-Y^2)/\sqrt{X^2+Y^2}$ are both N(0, 1) distributed [Bansal et al. (1999)]. More generally, if $X \sim N(0,\sigma_1)$ and $Y \sim N(0,\sigma_2)$ are independent, then sign(X) $(X^2/\sigma_1 - Y^2/\sigma_2)/\sqrt{X^2+Y^2}$ is N(0, 1) [Reid (1987)]. Similarly, if X is normally distributed with PDF $\phi(x)$ and CDF $\Phi(x)$, then $Y = 2\phi(x)\Phi(ax)$ has a skew-normal distribution (Chattamvelli and Shanmugam 2021, Chap. 8). Birnbaum-Saunders distribution is a nonlinear transformation of Gaussian variate as

$$X = \beta[\alpha Z/2 + \sqrt{(\alpha Z/2)^2 + 1}]^2, \quad (2.60)$$

where $Z \sim N(0, 1)$, α and β are positive constants ([Chattamvelli and Shanmugam (2022)]) Chap. 11). Distribution of sum of products of correlated bivariate-normal random vectors

$Z=\sum X_k Y_k$ finds applications in electrical engineering, astrophysics and quantum cosmology [Gaunt (2022)].

2.14 Applications

Distribution of a function of RV(s) has many practical applications. For instance, the sum of RVs is used to investigate the performance of communication systems [Du et al. (2020)]. Sum of RVs is used in DSP to find the effect of an impulse of magnitude k applied at time t on a signal, which results in the same signal scaled by the impulse magnitude and shifted to the location of the impulse as k f(x–t). The sum of IID uniform (rectangular) distributions is used in instrumentation (modeling uncertainties in parameters, calibration of measurement devices), numerical analysis and error modeling (estimating aggregated errors). The distribution of the sum of n IID rectangular variates (U(0, 1)) find applications in round-off error analysis, aggregating errors, etc. [Lusk and Wright (1982)]. It is called Irwin-Hall distribution (IHD). It is easy to obtain the PDF using the MGF technique as

$$f(x; n) = \frac{1}{\Gamma(n)} \sum_{k=0}^{n} (-1)^k \binom{n}{k} (x - k)_+^{n-1} \qquad (2.61)$$

where $(x - k)_+ = (x - k)$ if $x \ge k$ and is zero otherwise.[1] Let T denote the sum for n IID U(0, 1) variates. Then for n = 2, T has a triangular distribution. But for n≥3, the peak smooths out quickly and the kurtosis decreases with mean = mode = n/2. The RV T approaches normality for increasing n ((T–n/2)/$\sqrt{n/12}$ →N(0, 1) as n→ ∞). A geometric derivation of IHD using the inclusion-exclusion principle can be found in [Marengo et al. (2017)].

Distribution of sums of RVs is used in communication systems where a message passes through several independent devices like routers or servers toward a destination.

Similarly, in chemical engineering a fluid is transported through several tanks (say all identical in every aspect) connected in series. The time that the fluid resides in a tank before it moves to the next tank can be modeled using a statistical law (such as exponential law) so that the total time in all the tanks in series is the sum of IID exponential distributions. If the tank capacities or outlet flow rates differ, the scale parameter of the exponential law will also differ, so that we have a sum of exponential distributions with different scale parameters. Another example is the total amount of a chemical product (as in chemical and pharmaceutical industries) in fixed time t, so that the grand amount in an integer multiple of t (T = nt) can be found as a sum of RVs.

Sum of Pareto distributions is used in insurance (ruin theory and reinsurance pricing) [Morales (2005)], banking (cash-flow analysis) and finance (catastrophe bonds (also called CAT bonds)) [Nadarajah et al. (2017)], economics (income studies), geology (mineral deposits), geophysics and seismology (energy released in an earthquake or aftershocks)

---

[1] https://en.wikipedia.org/wiki/Irwin%E2%80%93Hall_distribution.

[Zaliapin et al. (2005)], radars (fuzzy target detection; adaptive detection of signals of unknown frequency), sonars (track initialization for multi-static sonar systems) [Hempel (2007)], petroleum engineering (mineral exploration or oil and gas prospecting), stock market (portfolio's aggregate losses) and waiting time distributions [Ramsay (2008)] etc. Sum of independent non-identically distributed RVs find applications in change-point analysis [Mitra (1971); Sadooghi-Alvandi et al. (2009)], and in aggregating scaled values with differing numbers of significant digits [Buonocore et al. (2009)]. A characterisation of this sum is discussed in [Mao et al. (2019)].

The sum of Pareto distributions is used to model loss-ratios (say due to catastrophic events) [Ghitany et al. (2021)] and insurance claim amounts exceeding a set threshold in a fixed time period (for Pareto claim size distributions) [Ramsay (2008)]. Compound sums of Pareto distributions are used in actuarial sciences [Ramsay (2009)]. Sums of truncated Pareto distributions are used in geology, atmospheric sciences[2] and hydrology (in which log-normal and gamma distributions are also used) [Genc (2021)]. Some random phenomena exhibits periodicity over a fixed time interval. Examples are precipitation levels, temperature, air quality index, etc. in a city which varies continuously over a 24 h interval.

The annual maximum flood occurrence dates (AMFOD) in hydrology uses von Mises distribution with PDF

$$f(x; \mu, k) = \exp(k \cos(x - \mu))/(2\pi I_0(k)), \ 0 < x \le 2\pi, \ 0 < \mu \le 2\pi, \qquad (2.62)$$

where μ is the mean position parameter, k is the concentration parameter, and $I_0(k)$ is the modified Bessel function of order zero. The annual maximum flood magnitude (AMFM) is modeled using displaced gamma law

$$f(x; c, \alpha, \beta) = (\beta^\alpha / \Gamma(\alpha)) \exp(-\beta(x - c))(x - c)^{\alpha-1}, \ x > c. \qquad (2.63)$$

The minimum of several IID RVs occurs in reliability analysis to assess the probability of failure or malfunction of devices or components. See [Gianfranco et al. (1995)] for the minimum of geometric and exponential RVs.

The distribution of difference (DoD) of two RVs appears in several fields. It is used in the analysis of space shift keying (SSK) based satellite communication systems [Kumar et al. (2021)].

Negative exponential mixtures are used to model manufacturing systems. The DoD of supply and demand is used in marketing, energy demand and total energy production in power engineering, crop yield, and the DoD of two independent Poisson distributions (called Skellam distribution) is used in game theory and two-person games. See [Joarder and Omar (2022)] for the distribution of difference of two chi-square RVs.

---

[2] Geological properties such as porosity, permeability, density, or mineral composition are often modeled as random fields over a spatial domain. Atmospheric properties are modeled in 3D or 4D where fourth dimension is time.

Distribution of transcendental functions has applications in mining engineering, reliability, spectroscopy, and communications, among many other fields. Dependent and nonidentical RVs are also common in some fields [Cahillane (2020)]. One example is in digital communications where the signals are often affected by uncertainties and noise. Amplitudes and phases of the noise components in signal processing applications are modeled using nonidentical RVs for which the copula method (Chap. 3, Sect. 3.7) is more suitable. Reliability analysis and risk assessment for electrical systems also use nonidentical RVs.

In data compression and telecommunications, the speech amplitude is modeled using the Laplacian law $f(x) = (1/\sigma\sqrt{2})\exp(-\sqrt{2}|x|/\sigma)$, and compressed using the $\mu-$law as $A*\text{sign}(x)[\ln(1+\mu|x|)/\ln(1+\mu)]$ where "A" is the peak input magnitude, and μ is the compression constant (typically set to high, say 255). Similarly, in wireless communication of fading channels, if X is $N(\mu, \sigma^2)$ distributed, $Y = e^X$ has a lognormal distribution.

Product of RVs find applications in portfolio diversification models [Nadarajah and Kotz (2007)]. Product and ratio of Rician RVs are used to identify the average fade duration of wireless communication systems and co-channel interference operating over Rician multipath fading channels [Krstic et al. (2016)]. The distribution of the product of two RVs in which one of them is the standard normal and the other is one among nearly fifty distributions is investigated in [Jiang and Nadarajah (2019)].

Product of triangular distributions is used in oil-exploration (the total yield from an oil field; one for the number of oil deposits and the other for the yield from each deposit) and nuclear waste management [Glickman and Xu (2008)].

Consider the electrical resistance of semiconductors, which depends on the temperature in °K as

$$r(t) = (a/t^b)e^{c/t} \quad \text{where t = temperature in K°,} \quad a,b,c \text{ are constants.} \tag{2.64}$$

If the temperature variation is known, the resistance distribution can be derived for some values of the parameters. Similarly, the potential energy of weakly interacting dielectric gas subjected to an external electric field is modeled using an exponential law as $f(y) = \exp(-y/(BT))$ where y = potential energy, B = Boltzmann constant, T = absolute temperature in °K. See [Khuong and Kong (2006)] for the general expression for the sum of n independent exponential RVs.

The distribution of F(x) and its inverse are encountered in several applications in reliability, survival analysis, income inequality analysis in economics [Shanmugam (1991)].

Ratio of RVs is used in stress-strength model of reliability, and many other fields. Ratio of RVs (Rayleigh, Nakagami-m, Weibull) finds applications in the analysis of multi-hop wireless transmissions [Mekic et al. (2012)]. The fatigue life of mechanical components subjected to cyclic loading are modeled using Birnbaum-Saunders distribution, which is a nonlinearly transformed normal RV [Chattamvelli and Shanmugam (2022)].

2.15   Summary

This chapter derives and explains the formulas for the probability distribution of a sum, difference, product and ratio of two independent RVs. The distribution of squares, square-roots, of univariate and other transformations of two or more RVs are derived and illustrated. Distribution of integer and fractional parts of some continuous RVs are discussed, as also the distribution of CDF (F(x)) and its inverse ($F^{-1}(x)$).

These known results are used to derive an expression for the mean deviation of some continuous distributions as twice the simple integral of $t/f(F^{-1}(t))$ from zero to $F(\mu)$ where μ is the mean and F(x) is the CDF ($F(\mu) = 1/2$ for symmetric laws). Several examples are included to understand the need and usefulness of the transformations.

2.16   Exercises

Problem 2.2 Mark as True or False

(a) The CoVT is applicable to both discrete and continuous distributions.
(b) If the range of a random variable X includes the origin, we cannot use the transformation Y = 1/X
(c) A translation Y = X±c is not applicable to discrete variates
(d) The distribution of the square of a normal variate is Student's T
(e) The reciprocal of a Cauchy variate is Cauchy distributed.

Problem 2.3 If X is $N(\mu, \sigma^2)$, find distribution of (i) Y = c*X + d, (ii) Y = $|X - \mu|$.

Problem 2.4 Find the constant c and $\Pr(x \geq 1/2)$ if f(x) = c x^2 for $|x| \leq 1$.

Problem 2.5 If f(x) = ke^{-kx}, x≥0, find the distribution of Y = $\sqrt{X}$ and Z = X^2 for k > 0.

Problem 2.6 If X is GEOM(p), find the distribution of Z = $\sum_{i=1}^{n} X_i$ where each X_i are independent.

Problem 2.7 If X is DUNI(k) with f(x) = 1/k, x = 1, 2, ..., k, find distribution of Y = X + b.

Problem 2.8 If X is U(0, 1), prove that Y = $\frac{1}{\lambda}\ln(1/X) = -\ln(X)/\lambda$ is exponentially distributed where ln denotes log to base e.

Problem 2.9 If X has PDF f(x), prove that the PDF of Y = cX is $(1/|c|)$ f(x/c).

Problem 2.10 If $\Phi(x)$ denotes the CDF of a normal distribution, prove that $\frac{1}{\Phi(a/\sqrt{a^2+b^2})}$ $\Phi(a+bx)\phi(x)$ is a well-defined random variable, which is independent of b when a = 0. Prove that its mean is $b/\sqrt{1+b^2}$.

Problem 2.11 If n is an integer ≥ 1, check whether $f(x) = \frac{1}{n^{3/2}(2\pi)^{(n-1)/2}}x^2[\phi(x)]^n$, $-\infty < x < \infty$ is a well-defined random variable.

Problem 2.12 If X is U(0, 1) prove that $Y = -\ln(X)$ is standard exponential. What is the distribution when the logarithm is not to the base e.

Problem 2.13 If h(x) is a monotonic function, prove that the CDF of $Y = h(x)$ can be expressed as $F_Y(y) = F_X(h^{-1}(y))$ if x is increasing, and $F_Y(y) = 1 - F_X(h^{-1}(y))$ otherwise.

Problem 2.14 If X~ CUNI(–b, +b) has a uniform distribution, find the distribution of $Y = \exp(aX)$.

Problem 2.15 If X is U(0, 1), find the distribution of $Y = 1-e^{-x}$.

Problem 2.16 If X is geometric with parameter p, find $\Pr[X^2 \leq 16]$.

Problem 2.17 If X_1, X_2 are IID U(0, 1), find the distribution of $Y_1 = \sqrt{-2\log_e(x_1)}$ $\cos(2\pi x_2)$ and $Y_2=\sqrt{-2\log_e(x_1)}\sin(2\pi x_2)$ (the inverse transformation being $X_1 = \exp\left[-\frac{1}{2}(y_1^2 + y_2^2)\right]$ and $x_2 = \frac{1}{2\pi}\arctan(y_2/y_1)$).

Problem 2.18 If $f(x) = 2x^2(x + c)$, for $0\leq x <1$, find the unknown c and the distribution of Y = 1/X.

Problem 2.19 Prove that the sum of independent exponential random variables has a gamma distribution.

Problem 2.20 If X $\sim$ BINO(n, p) conditional on p, where p is distributed as BETA(p, q), find the mean and variance of X.

Problem 2.21 If X is distributed as Gamma(α, 1), find the distribution of $Y = \log(X/\alpha)$.

Problem 2.22 Find the mean and variance of a random variable defined as $f(y) = \lambda/y^{1+\lambda}$, $1 \leq y < \infty, \lambda > 0$.

Problem 2.23 If $X_k \sim$ CUNI(-1/k,+1/k) where k is a nonzero number, find the distribution of $\sum_{k=1}^{n} |U_k|$ where n is an integer >1.

Problem 2.24 If $X \sim \chi_n$ (i.e.: $X \sim \sqrt{\chi_n^2}$) and $Y \sim$ BETA($\frac{n-1}{2}, \frac{n-1}{2}$) is independent of X, prove that $(2Y-1)X \sim$ N(0, 1).

Problem 2.25 Find the marginal distributions of X and Y if $f(x,y) = 1/(\pi r^2)$ for $x^2 + y^2 \le r^2$ where r>0.

Problem 2.26 If $D = (x, y)$ is a point on the interior of a circle centered at (0,0) prove that the expected value of the distance of D from the center is 2r/3 where r is the radius.

Problem 2.27 If X is a standard normal variate N(0, 1) and Y|X is distributed as $f(y|x) = (u/2)\exp(-u|y|)$ where u = $(1 + x^2)/2$, find (i) the unconditional distribution of Y, (ii) Show that $\sigma_{y|x}^2 = 8/(1 + x^2)^2$.

Problem 2.28 Find the distribution of absolute value of a general normal N(μ, σ^2), and its mean.

Problem 2.29 If $X \sim$ Pareto(a, b) with PDF f(x; a, b) = ab^a/x^{a+1} for x>b, find the distribution of Y = $-\ln$(X/b).

Problem 2.30 If X is CUNI($-$(a + b)/2, (a + b)/2) variate, find the distribution of Y = |X|.

Problem 2.31 If X is CUNI(a, b), find the constants c and d such that cX+d $\sim$ U(0, 1).

Problem 2.32 If X and Y are IID Laplace random variables, prove that the distribution of the difference Z = X$-$Y is f(z) = (exp($-$z) + z exp($-$z))/4.

Problem 2.33 If U is U(0, 1) find the distribution of (i) $|U - \frac{1}{2}|$, (ii) 1/(1 + U).

Problem 2.34 If X is CUNI(0, b) find the distribution of Y = $\sqrt{X}$ and its mean.

Problem 2.35 If $X \sim$ Pareto(λ) with minimum x_m, prove that the distribution of Y = $\log(X/x_m)$ is EXP(λ).

Problem 2.36 If Y = g(x) is an arbitrary function of a discrete random variable, prove that the PMF of Y is $f(y) = \sum_{x \in g^{-1}(y)} f(x)$.

Problem 2.37 If $X \sim$ GAMA(m_1, p) and $Y \sim$ GAMA$(m_2, p + 0.50)$ are IID then prove that $Z=2\sqrt{XY} \sim$ GAMA$(\sqrt{m_1 m_2}, 2p))$.

Problem 2.38 Let X and Y denote the waiting time and consultation time of a patient to a clinic. If the joint distribution is $f(x, y) = Cy$, for $0 \le x \le y \le 90$ min, find (i) unknown C, (ii) $\Pr[X + Y \le 100 | X > 30]$.

Problem 2.39 If (x,y) denotes the Cartesian coordinates of a point inside a unit circle centered at the origin, with joint PDF $f(x, y) = (24/\pi)x^2 y^2$ for $x^2 + y^2 \le 1$ and is zero outside, find the marginal PDF of X and check whether X and Y are independent.

Problem 2.40 Let (r, θ) denote the Polar coordinates of a point inside a unit circle centered at the origin (r is the distance of the point from the origin and θ is the angle made with the X-axis), find the joint distribution and check whether r and θ are independently distributed.

Problem 2.41 If X is standard Cauchy distributed, find the distribution of $Y = X^2$ and its mean.

Problem 2.42 The horizontal distance traveled by a projectile is given by $D = v^2 \sin(2\theta)/g$ where g is the acceleration due to gravity ($g \simeq 9.8$ ms^2), θ is the firing angle (with horizontal), and v is the initial velocity of a projectile. Find the PDF of D when $\theta \sim$ CUNI$(0, \pi/4)$.

References

Bansal, N., Hamedani, G.G. et al. (1999). Some characterizations of the normal distribution, *Stat. Prob. Lett.*, 42, 393–400. https://www.sciencedirect.com/science/article/pii/S0167715298002351, https://doi.org/10.1016/S0167-7152(98)00235-1

Buonocore A., Pirozzi E., Caputo, L. (2009). A note on the sum of uniform random variables, *Stat. and Prob. Lett.*, 79(19), 2092–2097, https://doi.org/10.1016/j.spl.2009.06.020

Cahillane, C. (2020). Sum of exponential and Laplace distributions, https://ccahilla.github.io/sum_exponential_and_laplace_distributions.pdf

Chattamvelli, R. & Shanmugam, R. (2020) Discrete Distributions in Engineering and the Applied Sciences, Springer. https://link.springer.com/book/10.1007/978-3-031-02425-2

Chattamvelli, R. & Shanmugam, R. (2021) Continuous Distributions in Engineering and the Applied Sciences—Part 1, Springer. https://doi.org/10.1007/978-3-031-02430-6

Chattamvelli, R. & Shanmugam, R. (2022) Continuous Distributions in Engineering and the Applied Sciences – Part 2, Springer. https://link.springer.com/book/10.1007/978-3-031-02435-1

Du, H., et al. (2020). Sum of Fisher-Snedecor F random variables and its applications, *IEEE Open J. of the Commu. Soc.*, 1, 342-356, https://ieeexplore.ieee.org/document/9044870, https://doi.org/10.1109/OJCOMS.2020.2982770.

Gaunt, R.E. (2022). The basic distributional theory for the product of zero-mean correlated normal random variables, arXiv:2106.02897v2

Genc, A.I. (2021). Products, sums and quotients of upper truncated Pareto random variables with an application in hydrology, *Water resource management*, 35(1), 369–383. https://link.springer.com/article/10.1007/s11269-020-02740-z

Ghitany, M.E., Gomez-Denis, E., Nadarajah, S. (2018). A new generalization of the Pareto distribution and its application to insurance data, *J. Risk Financial Manag.*, 11(1), https://www.mdpi.com/1911-8074/11/1/10, https://doi.org/10.3390/jrfm11010010

Gianfranco, C., Lawrence,L & David, N. (1995). On the minimum of independent geometrically distributed random variables, *Stat. and Prob. Lett.*, 23(4), 313–326, https://doi.org/10.1016/0167-7152(94)00130-Z

Glickman, T.S., Xu, F.(2008). The distribution of the product of two triangular random variables, *Stat. and Prob. Letters*, 78(16), 2821–2826, https://doi.org/10.1016/j.spl.2008.03.031

Hempel, C. G. (2007). Track initialization for multi-static active sonar systems, OCEANS 2007 - Europe, 1–6, https://ieeexplore.ieee.org/document/4302458

Jiang, X. & Nadarajah, S. (2019). On characteristic functions of products of two random variables, *Wireless Personal Commu.* https://doi.org/10.1007/s11277-019-06462-3

Joarder, A.H. & Omar, M.H. (2022) On the exact distribution of the difference between two chi-square variables *The Bull. of the Malaysian Math. Soc.*, Ser. 2, 45(38), 1–18, https://doi.org/10.1007/s40840-022-01373-2

Krstic, D., et al. (2016). Statistical characteristic of ratio and product of Rician random variables and its application in analysis of wireless communication systems, *Intl. J. of Math. and Comput. Methods*, 1, 79–86. http://www.iaras.org/iaras/journals/ijmcm

Khuong, H. V. & Kong, H. Y. (2006). General expression for pdf of a sum of independent exponential random variables, *IEEE Commu. Lett.*, 10, 159-161, https://ieeexplore.ieee.org/document/1603370

Kumar, M., Hazra, S., Arti, M.K. (2021) On the Distribution of the Difference of Two SR RV's and it's Application in Satellite Communication, 2021 8th *Intl. Conf. on Signal Proc. and Integrated Networks* (SPIN), Noida, India, 2021, pp. 1098-1103, https://doi.org/10.1109/SPIN52536.2021.9566110.

Lusk, E.J. and Wright, H. (1982). Deriving the probability density for sums of Uniform random variables, *The Amer. Statn.*, 36(2), 128–130, https://doi.org/10.2307/2684025

Mao, T. et al. (2019) Sums of standard uniform random variables, *J. of Appl. Prob.*, 56(3), 918-936 https://doi.org/10.1017/jpr.2019.52

Marengo, J.E., Farnsworth, D.L. and Stefanic, L. (2017). A geometric derivation of the Irwin-Hall distribution, https://www.hindawi.com/journals/ijmms/2017/3571419/, https://doi.org/10.1155/2017/3571419

Mekic, E., Mihajlo C. S., et al. (2012). Statistical analysis of ratio of random variables and its application in performance analysis of multihop wireless transmissions, *Math. Problems in Engg.*, 1–10. https://www.hindawi.com/journals/mpe/2012/841092/

Mitra, S.K. (1971). On the probability distribution of the sum of uniformly distributed random variables, *SIAM J. on Appl. Math.*, 20(2), 195–198, https://epubs.siam.org/doi/abs/10.1137/0120026, https://doi.org/10.1137/0120026

Morales, M. (2005). On an approximation for the surplus process using extreme value theory: Applications in ruin theory and reinsurance pricing, *N. American Actua. J.*, 8, 46–66

Nadarajah, S. & Kotz, S. (2007). On the product and ratio of t and Bessel random variables, *Bulletin of the Inst. of Maths.*, 2(1), 55–66. https://web.math.sinica.edu.tw/bulletin_ns/20071/2007103.pdf

Nadarajah, S., Zhang, Y., & Pogány, T. K. (2017). On sums of independent generalized Pareto random variables with applications to insurance and CAT bonds, *Probab. in the Engg. and Info. Sci.*, 1–10. https://doi.org/10.1017/S0269964817000055

Ramsay, C.M. (2008). The distribution of sums of IID Pareto random variables with arbitrary shape parameter, *Commu. Stat. - Theory and Meth.*, 37, 2177–2184. https://doi.org/10.1080/03610920500476325

Ramsay, C.M. (2009). The distribution of compound sums of Pareto distributed losses, *Scandi. Actu. J.*, 27–37.

Reid, J.G. (1987). Normal functions of normal random variables, *Comp. & Maths with Appl.*, 14(3), 157–160, https://www.sciencedirect.com/science/article/pii/0898122187901477, https://doi.org/10.1016/0898-1221(87)90147-7

Sadooghi-Alvandi, S.M., et al. (2009). On the distribution of the sum of independent uniform random variables, *Statistical papers*, 50(1), 171–175, https://link.springer.com/article/10.1007/s00362-007-0049-4, https://doi.org/10.1007/s00362-007-0049-4

Shanmugam, R. (1991). Significance testing of size bias in income data, *J. quant. econo.*, 7(2), 287–294.

Zaliapin, I.V., Kagan, Y.Y., Schoenberg, F. (2005). Approximating the distribution of Pareto sums, *Pure and Appl. Geophysics*, 162, 1187–1228. https://escholarship.org/uc/item/8940b4k8

Joint Distributions

3

This chapter introduces the distribution of functions of several random variables. The Jacobian of matrix transformation is described and its applications are cited. This is illustrated in finding the distribution of arbitrary transformations. Distribution of products and ratios of IID and dependent random variables are outlined. Plane-polar, spherical-polar, toroidal-polar and Helmert transformations are also discussed. A "do-little" technique to quickly find the Jacobian useful in statistics is described. Integration is heavily used in this chapter. The chapter ends with a discussion on Rosenblatt transforms, copula-based methods and applications of joint distributions.

3.1 Joint and Conditional Distributions

Joint probability distributions are used to model the simultaneous occurrence of multiple related events. For example the amount of rainfall received due to heavy thunderstorms or cloudbursts and subsequent river (water-level) exceedences are related events that can be modeled using statistical laws such as Weibull and Gumbel distributions which are popular in modeling the occurrence of extreme events. Similarly, joint probability distributions find applications in modeling uncertainties in building complex structures such as dams, nuclear reactors, highrises etc. and in stochastic processes.

Definition 3.1 Joint distribution is the distribution of two or more (dependent or independent) random variables.

© The Author(s), under exclusive license to Springer Nature Switzerland AG 2024   103
R. Chattamvelli and R. Shanmugam, *Random Variables for Scientists and Engineers*,
Synthesis Lectures on Engineering, Science, and Technology,
https://doi.org/10.1007/978-3-031-58931-7_3

Usually, the variables involved are either all discrete or all continuous. Symbolically, it is represented as p(x, y) = p(X = x and Y = y) (such that $\sum_{(x,y)\in A} p(x, y) = 1$ for discrete case, and $\iint_A p(x, y)dxdy = 1$ for the continuous case).

The joint CDF of two RVs uniquely define the cumulative event as

$$F_{XY}(x, y) = \text{Pr}(X \leq x, Y \leq y) = \begin{cases} \sum_{i=ll}^{x_1} \sum_{j=ll}^{y_1} p(x_i, y_j) & \text{if } X, Y \text{ discrete;} \\ \int_{ll}^{x_1} \int_{ll}^{y_1} p(x_i, y_j)dxdy & \text{if } X, Y \text{ continuous;} \end{cases}$$

where *ll* is the lower limit of the distribution. The subscripts (X, Y) will be dropped in the following discussions.

3.1.1   Properties of Bivariate CDF

The CDF satisfies the following properties.

1. $0 \leq F(x, y) \leq 1$.
2. F(−∞, −∞) = F(ll, ll) = 0; F(−∞, +∞) = F(ll, ul) = 0, F(ul, ul) = 1.
3. If $x_1 < x_2$ and $y_1 < y_2$ then F(x_1, x_2) < F(y_1, y_2).
4. $\partial^2 F(x, y)/[\partial x \partial y] = f(x, y)$ in continuous case, F(x, y) − F(x − 1, y) − F(x, y − 1) + F(x − 1, y − 1) = f(x, y) in discrete case.
5. $F_X(y) = F_{XY}(\infty, y)$, marginal CDF of Y, analogously $F_Y(x) = F_{XY}(x, \infty)$, marginal CDF of X.
6. $F_{XY}(-\infty, y) = 0$, $F_{XY}(x, -\infty) = 0$.
7. If X and Y are independent, F(x, y) = F(x) F(y).

where *ul* is the upper limit of the distribution. Notice that $F_X(-\infty, y) = 0$ (or $F_X(ll, y)$ in general) only in the limiting case. This may not hold in the neighborhood of *ll* when $\lim_{x+\to 0}$ (x approaches zero from right) for truncated distributions and those finite range distributions for which the mode is near the lower limit.

3.1.2   Marginal Distributions

Definition 3.2   Marginal distributions are distributions of individual variates.

Marginal PDF's can be obtained from joint PDF's by summation (in discrete case) or integration (in continuous case) as follows:–

$$f(x) = \sum_{y=-\infty}^{\infty} f(x, y), \quad \text{and } f(y) = \sum_{x=-\infty}^{\infty} f(x, y) \text{ (discrete),} \tag{3.1}$$

$$f(x) = \int_{y=-\infty}^{\infty} f(x, y)\, dy, \quad \text{and } f(y) = \int_{x=-\infty}^{\infty} f(x, y)\, dx \text{ (continuous)}, \qquad (3.2)$$

where the summation or integration is carried out only throughout the range of proper variate. Extension to more than two variates is straightforward. Joint PDF is the product of constituent marginal PDFs when the variables are independent.

$$f(x, y) = f(x)*f(y) \quad \text{and} \quad F(x, y) = F(x)*F(y) \qquad (3.3)$$

This has important applications in obtaining likelihoods, finding estimators, etc. We use PMF in the discrete case and PDF in the continuous case.

Find unknown K and marginal distribution

If $f(x, y) = K(x + y)$ for $0 < x < 1, 0 < y < 1$, find the constant K and obtain the marginal distributions.

Solution Consider $K\int_{x=0}^{1}\int_{y=0}^{1}(x + y)dxdy = K\int_{y=0}^{1}(x^2/2 + xy)|_0^1 dy = K\int_{y=0}^{1}(1/2 + y)dy$
$= K(y/2 + y^2/2)|_0^1 = K$ so that $K = 1$ and $f(x)$ becomes $f(x) = \int_{x=0}^{1}(x + y)dy = xy + y^2/2|_0^1$
$= (x + 1/2), 0 < x < 1$. Due to symmetry $f(y) = (y + 1/2), 0 < y < 1$.

Find marginal distribution

If the joint PMF of X and Y is given by $f(x, y) = K x(1 + y)$, $\{x = 1, 2\}$, $\{y = 1, 2, 3\}$, find the marginal PMF of X and Y.

Solution Both X and Y are discrete in this example. As the total probability is unity, we have $K[2 + 3 + 4 + 4 + 6 + 8] = 1$, giving $K = 1/27$. To obtain the marginal distribution of X, we sum out Y over its entire range. Hence $f(x) = (x/27)[9] = x/3$, $\{x = 1, 2\}$. Similarly, $f(y) = ((1 + y)/27)[3] = (1 + y)/9$, $\{y = 1, 2, 3\}$.

Radioactive α-particle receiver

A radioactive source is emitting α-particles intermittently in different directions. The number of particles emitted in a fixed time interval is Poisson(λ). A particle recorder is placed at a point in direct line-of-sight. It has probability p of recording any particle coming toward it. Find the PMF of the number of particles recorded.

Solution Let X be the number of particles emitted and Y be the number of particles recorded. Then we are given that

$$p(x) = \exp(-\lambda)\lambda^x/x!, \text{ and } p(y|x) = \binom{x}{y}p^y(1-p)^{x-y}. \tag{3.4}$$

As these two sources are independent, f(x, y) is the product of the individual PMFs. From this the marginal distribution of Y is obtained using (3.2) as

$$f(y) = \sum_x f(x, y) = \exp(-\lambda)p^y/y! \sum_{x=y}^{\infty} \lambda^x q^{x-y}/(x-y)!. \tag{3.5}$$

Put t = x − y in (3.5) so that t varies between 0 and ∞,

$$f(y) = \exp(-\lambda)(\lambda p)^y/y! \sum_{t=0}^{\infty} (\lambda q)^t/t! = \exp(-\lambda)(\lambda p)^y/y! * \exp(\lambda q). \tag{3.6}$$

This simplifies to $f(y, p, \lambda) = \exp(-\lambda p)(\lambda p)^y/y!$, for y = 0, 1, ..., as q = 1 − p.

3.1.3 Conditional Distributions

Conditional distributions are obtained from joint distributions by conditioning on one or more variables. Conditional PDF's can be expressed in terms of joint PDF's using laws of conditional probabilities.

$$f(x|y) = f(x, y)/f(y) \text{ and } f(y|x) = f(x, y)/f(x). \tag{3.7}$$

It is easy to see that multiple conditional distributions exist by conditioning y at different levels.

Gamma distributed Poisson parameter

Assume that the number of accidents follows a Poisson law with parameter λ. If λ itself is distributed according to the gamma law, prove that the unconditional distribution is negative binomial distributed.

Solution Let f(x, λ) represent the PMF of Poisson distribution and f(λ, m, p) denote the gamma PDF. Due to independence, the joint distribution is the product of the marginals and the unconditional distribution of x is obtained by integrating out λ as

$$f(x,m,p) = \int_{\lambda=0}^{\infty} [e^{-\lambda}\lambda^x]/x! * [m^p/\Gamma(p)]e^{-m\lambda}\lambda^{p-1}d\lambda. \tag{3.8}$$

Take constants independent of λ outside the integral to get

$$f(x,m,p) = m^p/[\Gamma(p)x!] \int_{\lambda=0}^{\infty} e^{-\lambda(m+1)}\lambda^{x+p-1}d\lambda. \tag{3.9}$$

The integral in (3.9) is easily seen to be the gamma integral, so that it becomes

$$m^p/[\Gamma(p)x!]\Gamma(x+p)/[(1+m)^{x+p}] = \binom{x+p-1}{x}(m/(1+m))^p(1/(1+m))^x,$$
(3.10)

where the last expression is obtained by writing $(1+m)^{x+p} = (1+m)^x * (1+m)^p$ and expanding the gamma functions as $\Gamma(x+p) = (x+p-1)!$ and $\Gamma(p) = (p-1)!$. This is a negative binomial distribution with $p = 1/(1+m)$.

Find Conditional distribution from trinomial

Let X and Y be jointly distributed as trinomial with PMF

$$f(x, y, n, p_1, p_2) = n!/[x!y!(n-x-y)!]p_1^x p_2^y(1-p_1-p_2)^{n-x-y}, \quad x+y \le n. \quad (3.11)$$

Find the conditional distribution of (i) $Y|X = x$, and (ii) $X| X+Y = n$. Obtain $E(Y|x)$, and $E(X|X+Y = n)$.

Solution We get the marginal PMF of y (resp x) by summing over x (resp y). Multiply and divide the RHS by $(n-y)!$ and sum over x to get

$$f(y) = \sum_x \frac{n!(n-y)!}{x!y!(n-y)!(n-x-y)!}p_1^x p_2^y(1-p_1-p_2)^{n-x-y}$$

$$= \frac{n!}{y!(n-y)!}p_2^y \sum_x \frac{(n-y)!}{x!(n-x-y)!}p_1^x(1-p_1-p_2)^{n-x-y}$$

$$= \binom{n}{y}p_2^y \sum_x \binom{n-y}{x}p_1^x(p_3)^{n-y-x} \quad (3.12)$$

where $p_3 = 1-p_1-p_2$. Expression inside the summation is simply the successive terms of the binomial expansion of $(p_1 + p_3)^{n-y}$. But, $(p_1 + p_3) = p_1 + (1-p_1-p_2) = (1-p_2)$. Substitute in the above to get $f(y) = \binom{n}{y}p_2^y(1-p_2)^{n-y}$, which is BINO$(n, p_2)$. Similarly, $X \sim$ BINO(n, p_1). The PMF of $Y|x$ is

$$\frac{f(x, y)}{f(x)} = \frac{(n-x)!}{y!(n-x-y)!}\left(\frac{p_2}{1-p_1}\right)^y\left(\frac{1-p_1-p_2}{1-p_1}\right)^{n-x-y}, \quad (3.13)$$

where $y = 0, 1, \ldots, n-x$. This is the PMF of BINO$(n-x, p_2/(1-p_1))$. Hence $E(Y|x) = (n-x)p_2/(1-p_1)$.

(ii) $X+Y$ is clearly distributed as a BINO$(n, p_1 + p_2)$, so that

$$P(X+Y = n) = \binom{n}{n}(p_1+p_2)^n(1-p_1-p_2)^{n-n} = (p_1+p_2)^n. \quad (3.14)$$

The PMF of $Y|X + Y = n$ is thus $f(x, n-x)/P[X + Y = n] = \frac{n!}{x!(n-x)!(n-x-(n-x))!} p_1^x p_2^{(n-x)} (1 - p_1 - p_2)^{n-x-(n-x)}/(p_1 + p_2)^n = \frac{n!}{x!(n-x)!} p_1^x p_2^{(n-x)}/(p_1 + p_2)^n$. Splitting $(p_1 + p_2)^n$ as $(p_1 + p_2)^x (p_1 + p_2)^{n-x}$, the above reduces to a BINO($n, p_1/(p_1 + p_2)$). From this we get $E(X|X + Y = n) = n p_1/(p_1 + p_2)$.

3.2   Jacobian Matrix

The Jacobian is a useful concept in various fields of applied sciences, including vector calculus, differential equations, atmospheric sciences, astronomy, econometrics, statistics, and
various branches of engineering. The Jacobian matrix is a matrix of first-order partial derivatives arranged as a rectangular (or square) array where the elements are usually functions
of input and output parameters. Carl Gustav Jacobi (1804–1851) whose work originated in
mathematical physics invented it in 1841. Small changes in the input parameters of a system
(such as concentrations, temperatures, or other process variables) may affect the overall
behavior of the system in process modeling, optimization, and control. This matrix contains
the partial derivatives of the output variables wrt the input variables in modeling problems
that involve many input and output variables (which need not tally in number). In other
words, the Jacobian relates infinitesimal areas in the input space to infinitesimal areas in
the output space of the same dimensionality (areas in 2–D, volume elements in $\geq$ 3–D). As
an example, digital communication systems use modulation techniques (such as amplitude
modulation, frequency modulation, or phase modulation) to share limited bandwidth for
multi-user data transfer. Small changes in the transmitted signal parameters (amplitude, frequency, or phase) may affect the received signal. Jacobian matrix can be used to analyze the
impact of the changes in system parameters, and for channel modeling. It is also useful in the
analysis of chemical reaction systems, transport phenomena, and other dynamic processes.
    It is used in mathematical statistics to map multiple variates or functions thereof to a new
and simplified space. The intension is either to separate-out common subsets of variables
or to make operations (such as integration) easier in the transformed space. The Jacobian
determinant measures the stretching effect of a mapping or transformation as explained later.
Jacobian could mean either the Jacobian matrix or it's determinant (if the matrix is square).
The Jacobian matrix could be rectangular when a mapping is induced from the Euclidean
space $\mathbb{R}^n \rightarrow \mathbb{R}^m$ where $m < n$. The impact or sensitivity on output variables due to a selected
subset of input variables (by keeping the other variables at fixed levels) can be studied by
the rows of the Jacobian matrix.
    As the determinant of a square matrix exists only if the matrix is of full-rank, there are
some regularity conditions to be satisfied by the transformations. We assume that there
are m real-valued functions $y_1 = f_1(x_1, x_2, \ldots, x_n)$, $y_2 = f_2(x_1, x_2, \ldots, x_n)$, $\ldots$, $y_m = f_m(x_1, x_2, \ldots, x_n)$. Then the Jacobian matrix comprises of all first-order partial derivatives
of mapping functions:–

$$J = \frac{\partial(x_1, x_2, \ldots, x_n)}{\partial(y_1, y_2, \ldots, y_m)} = \begin{bmatrix} \frac{\partial x_1}{\partial y_1} & \frac{\partial x_2}{\partial y_1} & \cdots & \frac{\partial x_n}{\partial y_1} \\ \frac{\partial x_1}{\partial y_2} & \frac{\partial x_2}{\partial y_2} & \cdots & \frac{\partial x_n}{\partial y_2} \\ \vdots & \vdots & \cdots & \vdots \\ \frac{\partial x_1}{\partial y_m} & \frac{\partial x_2}{\partial y_m} & \cdots & \frac{\partial x_n}{\partial y_m} \end{bmatrix}_{(m \times n)}.$$

The (i, j)th entry of the above matrix affirms that a small change dx_i in the original variate x should contract to $(\partial y_i / \partial x) dx_i$ in the transformed space. When m = n, the transformation is concisely expressed as the determinant of above matrix. The determinant of a square matrix $|J|$ is the same as the determinant of its transpose matrix $|J'|$. This means that the variable order is unimportant in statistical applications. A geometric interpretation of Jacobians is that it represents the best linear approximation to mapped domain at a general point using a tangent plane in the transformed space. Thus we get equivalent density contractions of space in transformed domain by multiplying the original function by the Jacobian (which acts as a magnification or contraction factor). For example

$$\iint_{R^2} f(x, y) dx dy = \iint_{R'^2} f(g(u, v), h(u, v)) \left| \frac{\partial(x, y)}{\partial(u, v)} \right| du dv. \tag{3.15}$$

Jacobian determinant is used to obtain the distribution of one-to-one (bijective) invertible functions of continuous random variables in statistics (the one-to-one condition can be relaxed in certain situations). Derivation is considerably simplified when the original variates are either statistically independent, or are identically distributed. All such transformation functions should be at least once differentiable. We denote the Jacobian determinant simply as $|J|$ (instead of $||J||$), where the vertical line has double meaning—it denotes the *absolute value* of the *determinant* (the vertical bar $|$ has various meanings in different fields—it denotes the absolute value of the argument in algebra, determinant in matrices, norm of a vector or a matrix in geometry, cardinality of a set or a set expression (like A∩B) in set-theory and probability theory, modulus in complex analysis and signal processing, etc. In some of the discussions below, the $|J|$ denotes only the determinant *without* absolute value.

3.3   Functions of Several Variables

Distribution of a function of random variable(s) has many applications in engineering and applied sciences. These are easily obtained when the variates are independent. It is fairly easy to obtain the joint distribution of identically distributed variables using a correct set of transformations. For functions of two variables, we have to choose a convenient auxiliary function such that the Jacobian is nonzero, and auxiliary variable is easy to integrate out (see Table 3.1).

The 2D Jacobian works on the principle that the average value of a double integral $f_{avg}(x, y) = 1/[(b - a) * (d - c)] \int_a^b \int_c^d dF(x, y)$ can be equalized under an arbitrary

Table 3.1 Common transformation of two variables

Transformation	Inverse transformation	Jacobian $|J|$
$u = x + y$, $v = x - y$	$x = (u + v)/2$, $y = (u - v)/2$	$-1/2$
$u = x \pm y$, $v = y$	$x = u \pm v$, $y = v$	1
$u = x + y$, $v = x/y$	$x = uv/(1 + v)$, $y = u/(1 + v)$	$-u/(1 + v)^2$
$u = x + y$, $v = x/(x + y)$	$x = uv$, $y = u(1 - v)$	$-u$
$u = ax + by$, $v = cx + dy$	$x = (du - bv)/D$, $y = (av - cu)/D$	$1/D$, $D = (ad - bc)$
$u = x/y$, $v = xy$	$x = \sqrt{uv}$, $y = \sqrt{v/u}$	$1/(2u)$
$u = mx/ny$, $v = y$	$x = nuv/m$, $y = v$	nv/m
$u = xy$, $v = y$	$x = u/v$, $y = v$	$1/v$
$u = \sqrt{x}$, $v = y$	$x = u^2$, $y = v$	$2u$
$u = \sqrt{xy}$, $v = y$	$x = u^2/v$, $y = v$	$2u/v$
$u = \sqrt{x + y}$, $v = \sqrt{y}$	$x = u^2 - v^2$, $y = v^2$	$4uv$
$u = \sqrt{x^2 + y^2}$, $v = y$	$x = \pm\sqrt{u^2 - v^2}$, $y = v$	$u/\sqrt{u^2 - v^2}$
$u = \sqrt{x^2 + y^2}$, $v = x/\sqrt{x^2 + y^2}$	$x = uv$, $y = u\sqrt{1 - v^2}$	$-u/\sqrt{1 - v^2}$
$u = \sqrt{nx}/\sqrt{y}$, $v = \sqrt{y}$	$x = uv/\sqrt{n}$, $y = v^2$	$2v^2/\sqrt{n}$
$u = x/y$, $v = xy/\sqrt{x^2 + y^2}$	$x = v\sqrt{u^2 + 1}$, $y = v\sqrt{u^2 + 1}/u$	$v(1 + 1/u^2)$
$u = xy/\sqrt{x^2 + y^2}$, $v = y$	$x = uv/\sqrt{v^2 - u^2}$, $y = v$	$v^3/(v^2 - u^2)^{3/2}$
$x = r\cos(\theta)$, $y = r\sin(\theta)$	$r = \sqrt{x^2 + y^2}$, $\theta = \tan^{-1}(y/x)$	r
$x = r\cos(\theta)\sin(\phi)$, $y = r\sin(\theta)\sin(\phi)$, $z = r\cos(\phi)$	$r = \sqrt{x^2 + y^2 + z^2}$, $\theta = \tan^{-1}(y/x)$	$-r^2\sin(\phi)$
$x = r\cos^2(\theta)$, $y = r\sin^2(\theta)$	$r = x + y$, $\theta = \tan^{-1}(\sqrt{y/x})$	$r\sin(2\theta)$

The order of the variables can be exchanged to get identical results, but the sign of Jacobian could differ. For example, if $u = xy$, $v = x$, $|J| = -1/v$. In the last case, as $x + y = r$, we can also write $\theta = \frac{1}{2}\cos^{-1}((x - y)/(x + y))$. The sign of $|J|$ is ignored in statistical applications. On line 5 $D \neq 0$

transformation $u = h(x, y)$; $v = g(x, y)$ as done in the univariate case. If the derivatives of $h(x, y)$ and $g(x, y)$ exist for each point in the rectangular region $[a, b] \times [c, d]$, then the above limit will be finite. This allows us to equate the width of an infinitesimal strip under the surface $f(x, y)$ as $f(x, y)dxdy = \gamma(u, v)dudv$ for all points within the region. From this we get $\gamma(u, v) = f(x, y)|J| = |J|f(h^{-1}(u, v), g^{-1}(u, v))$ where J is the Jacobian of the transformation. The only conditions in this transformation are that the mapping is once differentiable (ie. $h'(u, v)$, $g'(u, v)$) exists), and it is invertible (x, y can be expressed in terms of u, v) ($x = h^{-1}(u, v)$, $y = g^{-1}(u, v)$)). Note that the variables need not follow a statistical law in some applications such as image and video compression. Replace (x, y) by a vector of many variables to obtain the corresponding relationship in multivariate case.

Jacobian of U = $\frac{XY}{Z}$, V = $\frac{YZ}{X}$, and W = $\frac{ZX}{Y}$

If U = $\frac{XY}{Z}$, V = $\frac{YZ}{X}$, and W = $\frac{ZX}{Y}$, prove that the Jacobian is a constant.

Solution Consider UVW = $\frac{XY}{Z} * \frac{YZ}{X} * \frac{ZX}{Y}$ = XYZ, U/UVW = 1/VW = $1/Z^2$ gives Z^2 = VW or Z = $\sqrt{VW}$. Similarly V/UVW = 1/UW = $1/X^2$ gives X^2 = UW or X = $\sqrt{UW}$, Y = $\sqrt{UV}$. Jacobian is

$$|J| = \begin{vmatrix} w/2\sqrt{uw} & 0 & u/2\sqrt{uw} \\ v/2\sqrt{uv} & u/2\sqrt{uv} & 0 \\ 0 & w/2\sqrt{vw} & v/2\sqrt{vw} \end{vmatrix}$$

Take common factors outside the determinant to get

$$|J| = (1/8uvw) \begin{vmatrix} w & 0 & u \\ v & u & 0 \\ 0 & w & v \end{vmatrix}$$

= 2uvw/8uvw = 1/4, which is constant.

Find the Jacobian

Find the Jacobian of the transformation $y_1 = \sum_{i=1}^{n} X_i/\sqrt{n}$, $Y_2 = (X_1 - X_2)/\sqrt{2}$, $Y_i = (X_1 + X_2 + \cdots + X_{i-1} - (i-1)X_i)/\sqrt{i(i-1)}$ for i = 3, 4, ... n.

Solution From the first relation, we have $\sum_{i=1}^{n} X_i = \sqrt{n}y_1$. Put i = n to get $Y_n = (X_1 + X_2 + \cdots + X_{n-1} - (n-1)X_n)/\sqrt{n(n-1)}$. Add and subtract X_n on RHS and substitute for $\sum_{i=1}^{n} X_i = \sqrt{n}y_1$ to get $y_n = (\sqrt{n}y_1 - nx_n)/\sqrt{n(n-1)}$ so that $x_n = (\sqrt{n}y_1 - \sqrt{n(n-1)}y_n)/n$. From $Y_{n-1} = (X_1 + X_2 + \cdots + X_{n-2} - (n-2)X_{n-1})/\sqrt{(n-1)(n-2)}$ by adding and subtracting $X_{n-1} + X_n$ we get $y_{n-1} = (\sqrt{n}y_1 - (n-1)x_{n-1} - x_n)/\sqrt{(n-1)(n-2)}$. Substitute $x_n = (\sqrt{n}y_1 - \sqrt{n(n-1)}y_n)/n$ and separate out x_{n-1} to get $y_{n-1} = (\sqrt{n}y_1 + (\sqrt{n(n-1)}y_n - \sqrt{n}y_1)/n - (n-1)x_{n-1})/\sqrt{(n-1)(n-2)}$. Take $(n-1)x_{n-1}$ to the LHS and divide throughout by (n-1) to get $x_{n-1} = [\sqrt{n}y_1(1 - 1/n) + \sqrt{(1 - 1/n)}y_n - \sqrt{(n-1)(n-2)}y_{n-1}]/(n-1)$. Iterate similarly backwards to get subsequent values of x_i for i = n - 2, n - 3, ..., 1 to get the inverse mapping. First find partial derivatives to find the Jacobian. $\frac{\partial y_1}{\partial x_i} = 1/\sqrt{n}$ for i = 1,2, ...,n. $\frac{\partial y_2}{\partial x_1} = 1/\sqrt{2}$, $\frac{\partial y_2}{\partial x_2} = -1/\sqrt{2}$, $\frac{\partial y_2}{\partial x_i} = 0 \forall i > 2$.

$\frac{\partial y_3}{\partial x_1} = \frac{\partial y_3}{\partial x_2} = 1/\sqrt{6}, \frac{\partial y_3}{\partial x_3} = -2/\sqrt{6}, \frac{\partial y_3}{\partial x_i} = 0 \forall i > 3$, and so on. Jacobian is $|J| =$

$$
\begin{vmatrix}
1/\sqrt{n} & 1/\sqrt{n} & 1/\sqrt{n} & 1/\sqrt{n} & \cdots & 1/\sqrt{n} \\
1/\sqrt{2} & -1/\sqrt{2} & 0 & 0 & \cdots & 0 \\
1/\sqrt{6} & 1/\sqrt{6} & -2/\sqrt{6} & 0 & \cdots & 0 \\
1/\sqrt{12} & 1/\sqrt{12} & 1/\sqrt{12} & -3/\sqrt{12} & \cdots & 0 \\
\cdots & \cdots & \cdots & \cdots & \cdots & \cdots \\
\frac{1}{\sqrt{n(n-1)}} & \frac{1}{\sqrt{n(n-1)}} & \frac{1}{\sqrt{n(n-1)}} & \frac{1}{\sqrt{n(n-1)}} & \cdots & \frac{1-n}{\sqrt{n(n-1)}}
\end{vmatrix}.
$$

Many ways exist to evaluate this determinant as (i) expand it along the second row or last column (ii) take $1/\sqrt{n}$ as a common factor from first row and $1/\sqrt{n(n-1)}$ from last row and subtract first row from last row and expand remaining determinant from last column and last row in reverse. Thirdly, apply $C_2' = C_2 + C_1$ to get a zero in position $(2, 2)$, $2/\sqrt{6}$ at position $(3, 2)$ and so on. Next add the new second column C_2' to third column $(C_3' = C_2' + C_3)$ which reduces the element at position $(3, 3)$ to zero and so on to get $|J| =$

$$
\begin{vmatrix}
1/\sqrt{n} & 2/\sqrt{n} & 3/\sqrt{n} & 4/\sqrt{n} & \cdots & n/\sqrt{n} \\
1/\sqrt{2} & 0 & 0 & 0 & \cdots & 0 \\
1/\sqrt{6} & 2/\sqrt{6} & 0 & 0 & \cdots & 0 \\
1/\sqrt{12} & 2/\sqrt{12} & 3/\sqrt{12} & 0 & \cdots & 0 \\
\cdots & \cdots & \cdots & \cdots & \cdots & \vdots \\
\frac{1}{\sqrt{n(n-1)}} & \frac{2}{\sqrt{n(n-1)}} & \frac{3}{\sqrt{n(n-1)}} & \frac{4}{\sqrt{n(n-1)}} & \cdots & \frac{1-n}{\sqrt{n(n-1)}}
\end{vmatrix}.
$$

Expand along the second row repeatedly to get the determinant as 1.

Inverse mapping and Jacobian for u = x/(x + y), v = y/(x + y + z), w = x + y + z

Find the inverse mapping and Jacobian for the transformation u = x/(x + y), v = y/(x + y + z), w = x + y + z.

Solution Multiply v and w to get y = vw. Cross multiply u = x/(x + y) to get u(x + y) = x so that x(1 − u) = uy, or x = uy/(1 − u). Substitute for y to get x = uvw/(1 − u). Now x + y = vw + uvw/(1 − u) = vw[1 + u/(1 − u)] = vw/(1 − u), and z = w − (x + y) = w − vw/(1 − u) = w[1 − v/(1 − u)]. The Jacobian is

$$
|J| =
\begin{vmatrix}
vw/(1-u)^2 & uw/(1-u) & uv/(1-u) \\
0 & w & v \\
vw/(1-u)^2 & -w/(1-u) & 1 - u - v/(1-u)
\end{vmatrix}.
$$

Take common factors from column 1 and 2 to get

$$|J| = vw^2/(1-u)^2 \begin{vmatrix} 1 & u/(1-u) & -uv/(1-u) \\ 0 & 1 & v \\ 1 & -1/(1-u) & 1-u-v/(1-u) \end{vmatrix}.$$

This evaluates to $vw^2/(1-u)^2$.

Inverse mapping

Find the inverse mapping for the transformation $x = r\sin(\theta)\sin(\phi)$, $y = r\sin(\theta)\cos(\phi)$, $z = r\cos(\theta)\sin(\xi)$, $w = r\cos(\theta)\cos(\xi)$.

Solution Given $x = r\sin(\theta)\sin(\phi)$, $y = r\sin(\theta)\cos(\phi)$, $z = r\cos(\theta)\sin(\xi)$, $w = r\cos(\theta)\cos(\xi)$, so that

$$x^2 + y^2 = r^2\sin^2(\theta)[\sin^2(\phi) + \cos^2(\phi)] = r^2\sin^2(\theta), \tag{3.16}$$

$$z^2 + w^2 = r^2\cos^2(\theta)[\sin^2(\xi) + \cos^2(\xi)] = r^2\cos^2(\theta). \tag{3.17}$$

Add Eqs. (3.16) and (3.17) to get $x^2 + y^2 + z^2 + w^2 = r^2$ or $r = (x^2 + y^2 + z^2 + w^2)^{1/2}$.
Divide (3.16) by (3.17) to get $\tan^2(\theta) = (x^2 + y^2)/(z^2 + w^2)$ from which $\theta = \tan^{-1}((x^2 + y^2)/(z^2 + w^2))$.
Next take $x/y = \tan(\phi)$ from which $\phi = \tan^{-1}(x/y)$, and $z/w = \tan(\xi)$ so that $\xi = \tan^{-1}(z/w)$. The partial derivatives are as follows:
$\partial\xi/\partial x = \partial\xi/\partial y = 0$, $\partial\xi/\partial z = w/(w^2 + z^2)$, $\partial\xi/\partial w = z/(w^2 + z^2)$
$\partial\phi/\partial z = \partial\phi/\partial w = 0$, $\partial\phi/\partial x = y/(x^2 + y^2)$, $\partial\phi/\partial y = -x/(x^2 + y^2)$, $\partial r/\partial x = x/D$,
$\partial r/\partial y = y/D$, $\partial r/\partial z = z/D$, $\partial r/\partial w = w/D$ where $D = \sqrt{x^2 + y^2 + w^2 + z^2}$. From this the Jacobian can easily be found.

Distribution of difference of exponential RVs

If X and Y are IID exponential RVs with parameters λ_1 and λ_2, find the distribution of $U = X - Y$.

Solution Let $f(x;\lambda_k) = \lambda_k\exp(-\lambda_k x)$ be the PDFs for $k = 1, 2$ and $x \geq 0$. As they are independent, the joint PDF is $f(x, y) = K\exp(-(\lambda_1 x + \lambda_2 y))$ where $K = \lambda_1\lambda_2$. Take $U = X - Y$ and $V = Y$ so that $X = U + V$, and the Jacobian is 1. The joint PDF in terms of u and v is

$$f(u, v) = K\exp(-(\lambda_1(u + v) + \lambda_2 v)) = K\exp(-\lambda_1 u)\exp(-(\lambda_1 + \lambda_2)v). \tag{3.18}$$

As $U = X - Y$ can take positive and negative values, we need to consider two cases depending on $u < 0$ (ie. $x < y$) and $u > 0$ (i.e. $x > y$). In case (i), v varies from $-u$ to ∞ so that $f(u) =$

$$K \exp(-\lambda_1 u) \int_{v=-u}^{\infty} \exp(-(\lambda_1 + \lambda_2)v)dv = [K/(\lambda_1 + \lambda_2)] \exp(\lambda_1 u). \qquad (3.19)$$

For $u > 0$, integrate from 0 to ∞ to obtain $f(u) = [K/(\lambda_1 + \lambda_2)] \exp(-\lambda_2 u)$. Combine both cases to get

$$f(u) = [\lambda_1 \lambda_2/(\lambda_1 + \lambda_2)] \begin{cases} \exp(\lambda_1 u), & \text{for } u < 0; \\ \exp(-\lambda_2 u), & \text{for } u > 0 \end{cases}.$$

3.3.1 Arbitrary Transformations

The above transformation can be applied to arbitrary continuously differentiable, and invertible functions in higher dimensions (as bivariate, trivariate and multivariate transformations). Let x, y be jointly distributed according to some PDF f(x, y). Consider arbitrary continuously differentiable, and invertible functions of the form $u = g(x, y)$ and $v = h(x, y)$. If the mapping from (x, y) to (u, v) is one-to-one, it is invertible. We can express x and y in terms of u and v (say $x = G(u, v)$, $y = H(u, v)$). The differential relation f(x, y)dxdy = f(u, v) dudv translates into

$$f(u, v) = f(x, y) \begin{vmatrix} \frac{\partial x}{\partial u} & \frac{\partial x}{\partial v} \\ \frac{\partial y}{\partial u} & \frac{\partial y}{\partial v} \end{vmatrix} = f(G(u, v), H(u, v)) \begin{vmatrix} \frac{\partial x}{\partial u} & \frac{\partial x}{\partial v} \\ \frac{\partial y}{\partial u} & \frac{\partial y}{\partial v} \end{vmatrix}.$$

Here either u or v is the required transformation, and the other is called the *auxiliary* function. The choice of the auxiliary function is quite often arbitrary. It can be as simple as one of the original variables, provided that the inverse transformation is easy to find. It can be polar or trigonometric transformation when expressions like $\sqrt{x^2 + y^2}$ or $x^2 \pm y^2$ are present. Bivariate linear transformation is a special case in which the dependency is $u = c_1 x + c_2 y$ and $v = c_3 x + c_4 y$, where c_i's are arbitrary constants. The Jacobian of the transformation considerably simplifies in this case as

$$|J| = \begin{vmatrix} \frac{\partial(x, y)}{\partial(u, v)} \end{vmatrix} = \begin{vmatrix} c_1 & c_2 \\ c_3 & c_4 \end{vmatrix}.$$

A challenge in this type of transformations is the range mapping. We could visualize the transformed mapping easily in the bivariate case, but it is not easy in higher dimensions.

Do-little method to find Jacobian of Transformations:

Finding the determinant of a transformation involves much work in some cases. This can be reduced by the *do-little* method. Consider the transformation u = g(x, y) and v = h(x, y). We can reduce the work by taking v = y (or equivalently u = x) (this is called the identity mapping). Simply substitute for y = v in u to get u = g(x,v). Next find the derivative $\partial u/\partial x = \partial g(x,v)/\partial x$. Substitute for x (in terms of u and v) and take the reciprocal to get the Jacobian. Alternatively, express x = G(u, v) and find J $= \partial x/\partial u = \partial G(u, v)/\partial u$. As examples, consider the transformation u = x + y, v = y. Put y = v to get u = x+v, and 1/J = $\partial u/\partial x$ =1; as v is constant. Alternatively, solve for x to get x = u-y = u-v. Then J = $\partial x/\partial u = (\partial/\partial u)(u - v) = 1$. Similarly, if u = xy, v = y then x = u/v, and J = $\partial x/\partial u = 1/v$; and for u = x/y, v = y we have x = uv and J = $\partial x/\partial u$ = v. As another example, if u = x/(x + y) and v = y, x = uv/(1 – u) and J = $\partial x/\partial u$ = v/(1 – u)². This works even for constant multiples. If u = kx/y, v = y; we have x = uv/k and J = v/k.

    This idea can be extended to those cases where one of the input variables is a function of just one output variable in the two variables case. Consider u = $\sqrt{x + y}$ and v = $\sqrt{y}$ so that x = $u^2 - v^2$ and y = v^2. Then J = 2u*2v = 4uv. When the distribution of u = h(x, y) and v = g(x, y) are needed where both of them are complex functions of x and y, we can reduce the work with do-little method by doing this in two steps:– (i) use identity mapping on v as {u = h(x, y), v = x} and find the distribution of u with do-little method. (ii) use identity mapping on u as {u = x, v = h(x, y)} and find the distribution of v with do-little method. In multivariate transformations, we prudently choose our auxiliary functions in such a way that the resulting J matrix has zeros along lower or upper triangular positions (off main-diagonal) so that the determinant is simply the product of diagonal elements. As an example, let u = xy/z, v = (z + y)/(z – y) and w = z^2 so that z = $\sqrt{w}$, y = $\sqrt{w}$(v – 1)/(v + 1), and x = u(v + 1)/(v – 1). As y does not have u, and z does not have u and v, all of the off-diagonal elements of J matrix are zeros.

$$|J| = \begin{vmatrix} (v+1)/(v-1) & X & 0 \\ 0 & 2\sqrt{w}/(v+1)^2 & (v+1)/[2(v-1)\sqrt{w}] \\ 0 & 0 & 1/(2\sqrt{w}) \end{vmatrix} = \frac{1}{v^2-1}.$$

Distribution of sum of rectangular variates

An electronic circuit consists of two independent identical transistors connected in parallel. Let X and Y be the lifetimes of them, distributed as CUNI(0,b) with PDF f(x) = 1/b, 0 < x < b. Find the distribution of (i) Z = X + Y (ii) U = XY.

Solution Consider the transformation Z = X + Y, W = Y. The inverse transformation is Y = W, X = Z – W. The absolute value of the Jacobian is

$$|J| = \begin{vmatrix} -1 & 1 \\ 1 & 0 \end{vmatrix} = |-1| = 1.$$

Fig. 3.1 Region for Z = X + Y

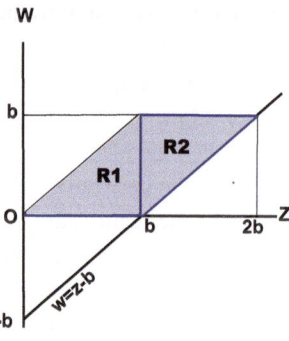

Fig. 3.2 Region for U = XY

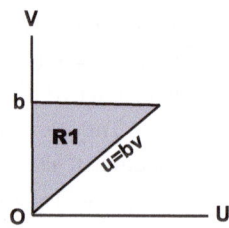

As X and Y are IID, the joint PDF is f(x, y) = $1/b^2$. The joint PDF of W and Z is f(w,z) = $1/b^2|J|$ = $1/b^2$. The range for Z is [0,+2b], and for W is [0,b]. As $0 < x < b$, we need to impose the condition $0 < z - w < b$. This in turn results in two regions of integration as shown in Fig. 3.1. For $0 < Z < b$, w varies between 0 and z so that f(z) = $\int_w f(w, z)dw = \int_{w=0}^{z} dw/b^2 = z/b^2$. For $b < z < 2b$, w varies between z-b and b, so that f(z) = $\int_w f(w, z)dw = \int_{w=z-b}^{b} dw/b^2 = $ (2b-z)/b^2. Combining both cases, we can write the PDF as f(z) = (b−|b − z|)/b^2 for 0<z<2b, because when z<b, |b − z| = b − z; and for z≥b, |b − z| = z-b. In the second case we put U = XY and V = Y so that X = U/V, and J = 1/v. The range for U is [0, b^2], and for V is [0, b]. As $0 < x < b$, we need to impose the condition 0<U/V<b, or equivalently u<bv. The region of integration is shown in Fig. 3.2. The joint PDF of u and v is f(u, v) = $1/(vb^2)$, $0 < u < bv < b^2$. The PDF of U is obtained as f(u) = $\int_{u/b}^{b} 1/(vb^2)dv = (1/b^2)(\ln(b)\text{-}\ln(u/b))$, 0<u< b^2. Using log(x/y) = log(x) − log(y), this can be simplified as $(1/b^2)[2 \ln(b)\text{-}\ln(u)]$.

Distribution of sum of exponential variates

If X and Y are IID exponentially distributed, find the distribution of X + Y and X − Y.

Solution This is most easily solved using the MGF method. We know that the MGF of X is $(1 − t/\lambda)^{-1}$. As X and Y are independent, $M_{x+y}(t) = M_x(t) * M_y(t) = (1 − t/\lambda)^{-2}$, which is the MGF of gamma($2,\lambda$). See [Cahillane (2020)] for an alternate derivation. Similarly, $M_{x-y}(t) = M_x(t) * M_y(-t) = (1 - t/\lambda)^{-1} * (1 + t/\lambda)^{-1} = (1 - t^2/\lambda^2)^{-1}$.

Take u = x + y, v = x − y so that $|J| = 1/2$, x = (u + v)/2, y = (u − v)/2. As they are independent, $f(x, y) = \lambda_1\lambda_2 \exp(-\lambda_1 x - \lambda_2 y)$. Substitute the values of x and y and multiply by the Jacobian to get $f(u, v) = \lambda_1\lambda_2 \exp(-\lambda_1(u + v)/2 - \lambda_2(u - v)/2)/2$. Integrate out v to get the PDF of U as

$$f(u; \lambda_1, \lambda_2) = (\lambda_1\lambda_2/2) \exp(-u(\lambda_1 + \lambda_2)/2) \int_v \exp(-v(\lambda_1 - \lambda_2)/2)dv. \qquad (3.20)$$

As v = x − y can take negative values and y = (u − v)/2 is positive, the range of v is $-u$ to u. Thus (3.20) reduces to $(\lambda_1\lambda_2/2) \exp(-u(\lambda_1 + \lambda_2)/2) * 2/(\lambda_1 - \lambda_2)[\exp(-u(\lambda_1 - \lambda_2)/2) \exp(u(\lambda_1 - \lambda_2))]$. The 2 cancels out giving the PDF of U as $f(u) = [\lambda_1\lambda_2/(\lambda_1 - \lambda_2)] [\exp(-u(\lambda_2)) - \exp(-u(\lambda_1))]$, which is called hypoexponential distribution.

Difference of rectangular variates

If X, Y are IID CUNI(0, b) distributed, find the distribution of (i) X − Y, (ii) |X-Y|.

Solution Make the transformation u = x − y, v = y so that x = u + v, $|J| = 1$. As X and Y are IID, $f(x, y) = 1/b^2$, so that $f(u, v) = 1/b^2$. Distribution of U is obtained by integrating out v as $f(u) = 1/b^2 \int_{u-b}^u dv = (b - u)/b^2$. In case (ii) consider u>0. As above $f(u, v) = 1/b^2$. We have to consider two cases $-b < u < 0$, and $0 < u < b$. As $0<u+v<b$, $f(u) = (1/b^2) \int_0^{b-u} dv = (b - u)/b^2$. From this the distribution of $|U - V|$ is obtained as

$$f(u) = \begin{cases} (u + 1)/b^2, & -b < u < 0; \\ (2 - u)/b^2, & 0 \le u < 1. \end{cases}$$

3.3.2   Image Jacobian Matrices

Image Jacobian matrices used in robotics, unmanned aerial vehicles (UAV), image and video compression, medical imaging etc are often rectangular. In image compression and video processing applications, we look for an unknown displacement vector (or matrix) to minimize two successive time frames (or sub-frames of appropriate sizes) so as to align successive images with minimal loss of information. Sparse residual distortions indicate almost still image frames. In this case the Jacobian matrix becomes

$$1/J = \frac{\partial(y_1 - u_{y_1}, \ldots, y_m - u_{y_m})}{\partial(x_1, x_2, \ldots, x_n)} = \begin{bmatrix} \frac{\partial y_1 - u_{y_1}}{\partial x_1} & \frac{\partial y_1 - u_{y_1}}{\partial x_2} & \cdots & \frac{\partial y_1 - u_{y_1}}{\partial x_n} \\ \frac{\partial y_2 - u_{y_2}}{\partial x_1} & \frac{\partial y_2 - u_{y_2}}{\partial x_2} & \cdots & \frac{\partial y_2 - u_{y_2}}{\partial x_n} \\ \vdots & \vdots & \cdots & \vdots \\ \frac{\partial y_m - u_{y_m}}{\partial x_1} & \frac{\partial y_m - u_{y_m}}{\partial x_2} & \cdots & \frac{\partial y_m - u_{y_m}}{\partial x_n} \end{bmatrix}$$

with the determinant sign (when the matrix is square, ie. m = n) indicating volumetric expansion ($|J| > 1$), shrinkage ($|J| < 1$) or steadiness ($|J| = 1$).

The Jacobian determinant is a function of the variates (or a constant) when applied to variate transformations in statistics (see Table 3.1). This means that the determinant can be made nonzero almost always. As the first derivative of linear functions is a constant, $|J|$ is a scalar constant for linear transformations (If Y = AX, then $|J| = |A|$ for multivariate transformation). Range of the transformed variates should be adjusted to account for this fact. As the Jacobian determinant is used as a multiplier, we take the absolute value of Jacobian in statistical applications (the sign of a determinant depends upon the order of the columns (variables) in the corresponding matrix). We can do better without the Jacobian method for simple transformation of variates like translations (u = x + c, v = y + d), and scaling (u = c*x, v = d*y) (using the CDF or MGF methods). Power of the Jacobian method becomes obvious when variable interactions are present.

This technique is applicable to discrete random variables as well. Let u = g(x, y) and v = h(x, y) be the bivariate mapping as before. Find the inverse transformation (express x and y as functions of u and v, say x = $f_1(u, v)$ and y = $f_2(u, v)$. Then the joint PMF of u and v is $p_{UV}(u, v) = p_{XY}(f_1(u, v), f_2(u, v))$. Now sum wrt u to get PMF of v and vice versa.

Functions of exponential distribution

Let X_i's be IID EXP(λ) with PDF f(x_i) = $\frac{1}{\lambda}e^{-x_i/\lambda}$. Define new variates Y_i's as $Y_1 = X_1/(X_1 + X_2 + \cdots + X_n)$, $Y_2 = (X_1 + X_2)/(X_1 + X_2 + \cdots + X_n)$, etc $Y_k = (X_1 + X_2 + \cdots + X_k)/(X_1 + X_2 + \cdots + X_n)$, and $Y_n = (X_1 + X_2 + \cdots + X_n)$. Prove that the joint distribution of $(Y_1, Y_2, \ldots, Y_n)$ depends upon y_n and y_{n-1} only.

Solution As X_i's are IID, the joint PDF is the product of individual PDFs. Thus f($x_1, x_2, \ldots, x_n$) = $\frac{1}{\lambda^n}e^{-\sum_{i=1}^n x_i/\lambda} = \frac{1}{\lambda^n}e^{-y_n/\lambda}$. The inverse mapping is $x_1 = y_1 y_n$, $x_2 = y_n(y_2 - y_1)$, $x_3 = y_n(y_3 - y_2)$, $x_k = y_n(y_k - y_{k-1}) \cdots, x_n = y_n(1 - y_{n-1})$. The Jacobian is

$$|J| = \left| \frac{\partial(x_1, x_2, \ldots, x_n)}{\partial(y_1, y_2, \ldots, y_n)} \right| =$$

$$
\begin{vmatrix}
y_n & 0 & 0 & 0 & \cdots & 0 & y_1 \\
-y_n & y_n & 0 & 0 & \cdots & 0 & y_2 - y_1 \\
0 & -y_n & y_n & 0 & \cdots & 0 & y_3 - y_2 \\
0 & 0 & -y_n & y_n & \cdots & 0 & y_4 - y_3 \\
\vdots & \vdots & \vdots & \vdots & \cdots & 0 & y_k - y_{k-1} \\
0 & 0 & 0 & 0 & \cdots & -y_n & 1 - y_{n-1}
\end{vmatrix}.
$$

To evaluate this determinant, we apply the row transformations $R'_2 = R_2 + R_1$, $R'_3 = R_3 + R'_2, \ldots, R'_{n-1} = R_{n-1} + R'_{n-2}$, and keep the nth row intact. The determinant reduces to

$$
\begin{vmatrix}
y_n & 0 & 0 & 0 & \cdots & 0 & y_1 \\
0 & y_n & 0 & 0 & \cdots & 0 & y_2 \\
0 & 0 & y_n & 0 & \cdots & 0 & y_3 \\
0 & 0 & 0 & y_n & \cdots & 0 & y_4 \\
\vdots & \vdots & \vdots & \vdots & \cdots & 0 & y_k \\
0 & 0 & 0 & 0 & \cdots & -y_n & 1 - y_{n-1}
\end{vmatrix}.
$$

By expanding this determinant along the first column, we get $|J| = y_n^n(1 - y_{n-1})$. Thus $f(y_1, y_2, \ldots, y_n) = \frac{1}{\lambda^n} e^{-y_n/\lambda} y_n^n (1 - y_{n-1})$, $0 < y_{n-1} < 1$, $y_n > 0$ which depends on y_{n-1} and y_n only.

3.3.3 Distribution of Products and Ratios

Product and ratio of IID random variables have many applications. Distribution of products of RVs can be found using Jacobian method, MoDF, multiple integration method and copula methods. These can be obtained by the Jacobian technique when the variables involved are continuous and independent. Make the transformation U = X/Y and V = XY. Then $x = \sqrt{uv}$, and $y = \sqrt{v/u}$, so that the Jacobian is $1/(2u)$. The joint PDF is the product of the marginal PDFs (due to independence assumption). From this the PDF of either of them can be obtained by integrating out the other. An alternate and simple method exists using the MoDF discussed in the last chapter.

Let F(z) be the CDF of the product. By definition,

$$
F(z) = P[Z \le z] = \int \int_{xy \le z} f(x, y)\,dx\,dy. \tag{3.21}
$$

As xy = c represents a parabolic curve, we split the range of integration of y from $(-\infty, z/x]$ and from $[z/x, \infty)$ to get

$$
F(z) = \int_{-\infty}^{0} \left[\int_{z/x}^{\infty} f(x, y)\,dy \right] dx + \int_{0}^{\infty} \left[\int_{-\infty}^{z/x} f(x, y)\,dy \right] dx. \tag{3.22}
$$

Using the transformation $U = XY$ this becomes

$$F(z) = \int_{-\infty}^{0} [\int_{z}^{-\infty} f(x, u/x) du/x + \int_{0}^{\infty} [\int_{-\infty}^{z} f(x, u/x) du/x] dx. \qquad (3.23)$$

This upon rearrangement becomes

$$F(z) = \int_{-\infty}^{z} [\int_{-\infty}^{\infty} (1/|x|) f(x, u/x) dx] du. \qquad (3.24)$$

Differentiate wrt z to get the PDF as

$$f(z) = \int_{-\infty}^{\infty} (1/|x|) f(x, z/x) dx. \qquad (3.25)$$

It is shown below that if X and Y are IID normal variates, the ratio X/Y has a Cauchy distribution. Analogously $U = X/Y$ has PDF

$$f(u) = \int_{-\infty}^{\infty} |x| f(x, ux) dx. \qquad (3.26)$$

The ratio of RVs may be written as a product and the above technique used to find the PDF. As an example, $U = X/Y$ can be written as $U = X \times (1/Y)$, so that the distribution of U is the product of the distributions of X and 1/Y.

Product of Rectangular distributions

What is the Jacobian of the transformation $u = x(1 - y)$, $v = xy$. Use it to find the distribution of u and v when x and y are U(0, 1).

Solution Take the ratio to get $U/V = (1 - Y)/Y$. This gives $Y = U/(U + V)$, $X = U + V$. The Jacobian is

$$|J| = \left| \begin{matrix} 1 & 1 \\ -v/(u + v)^2 & u/(u + v)^2 \end{matrix} \right|$$

This simplifies to $1/(u + v)$. If X, Y are IID U[0, 1], f(x, y) = 1 so that f(u, v) = 1/(u + v). As v ranges from 0 to 1 – u, we get the PDF of U as $f(u) = \int_{v=0}^{1-u} dv/(u + v) = \log(u + v)|_0^{1-u} = -\log(u)$ for $0 < u < 1$.

Log of ratio of U(0, 1) distribution

If x, y are IID U(0, 1), find the PDF of (i) U = -log(X/Y), (ii) V = X^2.

Solution We have f(x, y) = 1. Write U = –log(X/Y) = log(Y) – log(X) = V – W (say). Distribution of V = log(Y) can easily shown to be the standard exponential using MoDF. Hence the problem reduces to finding the distribution of the difference of two IID exponential

Fig. 3.3 Region of integration $Y = X^2$

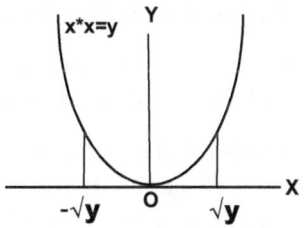

RVs (this is already considered in Sect. 3.3). Write $u = v - w$, $z = w$ to get $v = u + z$. The Jacobian of the transformation is 1 (Table 3.1). As $f(v, w) = \exp(-(v + w))$, the joint distribution of u and z is $f(u, z) = \exp(-(u + 2z))$. The distribution of u is obtained by integrating out z. As U can take positive and negative values, we split the integral into two parts (for $u < 0$ and $u > 0$). For $u < 0$, z must be greater than $-u$ (as $v > 0$), so that $f(u) = \int_z f(u, z)dz = \exp(-u)\int_{-u}^{\infty}\exp(-2z)dz = \exp(u)/2$, for $u \leq 0$. For $u > 0$ consider $\Pr(U \leq u) = 1 - \Pr(U > u)$. This is $1 - \Pr[W \geq V - u]$. Proceeding as above gives $\Pr[W \geq V - u] = \exp(-u)\int_0^{\infty}\exp(-2z)dz$, so that the PDF becomes $f(u) = 1 - \exp(-u)/2$. Combining both cases shows that it is the PDF of Laplace distribution.[1]

In part (ii), the easiest way to find the PDF of $V = X^2$ is using CDF method (Fig. 3.3) as $F(v) =$

$$P(X^2 \leq v) = P(-\sqrt{v} \leq X \leq \sqrt{v}) = F(\sqrt{v}) - F(-\sqrt{v}) = \frac{1}{2b} * 2\sqrt{v} = \sqrt{v}/b. \quad (3.27)$$

From this the PDF is obtained by differentiation as $f(v) = 1/(2b\sqrt{v})$

Conditional distribution of X|Y

If $f(x, y) = \Gamma(m)/[\Gamma(a)\Gamma(b)\Gamma(m - a - b - 1)]\,x^{a-1}y^{b-1}(1 - x - y)^{m-a-b-1}$, prove that the marginals and conditional distributions of X|Y, Y|X are all beta distributed when the variables are independent.

Solution Obviously $0 \leq x$, $y \leq 1$, and $y \leq 1 - x$. Consider $f(x) = kx^{a-1}\int_{y=0}^{1-x}y^{b-1}(1 - x - y)^c dy$ where $c = m - a - b - 1$. Put $y - (1 - x)t$ so that $dy = (1 - x)dt$. When $y = 0$, $t = 0$ and when $y = 1 - x$, $t = 1$. As $dy = (1 - x)dt$, $f(y) = kx^{a-1}(1 - x)^{c+1}\int_0^1 t^{b-1}(1 - t)^c dt$. The integral is the complete beta function $B(b, c+1)$ so that Y is beta distributed.

---

[1] Note that the difference of two independent exponential distributions with parameters λ_1 and λ_2 is an asymmetric Laplace distribution, which becomes standard Laplace distribution when $\lambda_1 = \lambda_2$.

Ratio of Cauchy distributions

If X and Y are independent Cauchy distributed, prove that (i) Z = (X + Y)/2, (ii) (X − Y)/(1 + XY) are identically distributed (iii) X/Y and XY are both identically distributed.

Solution We know that the ChF of Cauchy distribution is exp(−|it|). To find the ChF of Z = (X + Y)/2, first use $M_{cx}(t) = M_x(tc)$ where c = 1/2 and then use $M_{x+y}(t) = M_x(t) * M_y(t)$ to get $M_z(t) = \exp(-|it/2|) * \exp(-|it/2|) = \exp(-|it|)$ which is the ChF of Cauchy distribution. For part (ii) put X = tan(U), Y = tan(V) so that (X − Y)/(1 + XY) = [tan(U)− tan(V)]/[1 + tan(U) tan(V)] = tan(U − V). As X = tan(U), U = $\tan^{-1}(X)$ is uniform distributed (Chap 2 example Sect. 2.11.1), and so is V = $\tan^{-1}(Y)$. (iii) As the Cauchy distribution is the ratio of two IID normal distributions, we will define $C_1 = Z_1/Z_2$ and $C_2 = Z_3/Z_4$ where each of the Z_k's are independent N(0, 1). Consider $C_1/C_2 = (Z_1/Z_2)/(Z_3/Z_4)$. Write this as $C_1/C_2 = (Z_1/Z_2) \times (Z_4/Z_3)$. This shows that product and ratio of IID Cauchy variates are identically distributed.

Ratio of independent uniform distributions

If X and Y are IID CUNI(0, b) variates, find the distribution of U = X/Y.

Solution Let U = X/Y, V = Y so that the inverse mapping is Y = V, X = UV. The Jacobian is |J| = v. The joint PDF is f(x, y) = $1/b^2$. Hence f(u, v) = v/b^2. The PDF of u is obtained by integrating out v. A plot of the mapping is shown in Fig. 3.2. The region of interest is a rectangle of sides 1xb at the left, and a curve uv = b to its right. Integrating out v, we obtain f(u) = $\int_0^b \frac{v}{b^2} dv$ for 0< u ≤1, and f(u) = $\int_0^{b/u} v/b^2 dv = 1/(2u^2)$ for 1< u < ∞.

$$
f(u) = \begin{cases} 1/2 & \text{for } 0 < u < 1; \\ 1/(2u^2) & \text{for } 1 < u < \infty \end{cases}
$$

which is independent of the parameter b.

Sum and ratio of independent gamma distributions

If X and Y are IID GAMMA(α, β_i), find the distribution of (i) X + Y, (ii) X/Y.

Solution Let U = X + Y and V = X/Y. Solving for X and Y in terms of U and V, we get x = $\frac{uv}{1+v}$, and y = $\frac{u}{1+v}$. The Jacobian of the transformation is

$$
|J| = \begin{vmatrix} \frac{v}{1+v} & \frac{u}{(1+v)^2} \\ \frac{1}{1+v} & -\frac{u}{(1+v)^2} \end{vmatrix} = \frac{-u}{(1+v)^2}.
$$

The joint PDF of X and Y is f(x, y) = $\frac{\alpha^{\beta_1+\beta_2}}{\Gamma(\beta_1)\Gamma(\beta_2)} e^{-\alpha(x+y)} x^{\beta_1-1} y^{\beta_2-1}$. Multiply by the Jacobian, and substitute for x, y to get

$$f(u, v) = \frac{\alpha^{\beta_1+\beta_2}}{\Gamma(\beta_1)\Gamma(\beta_2)} e^{-\alpha u} (uv/(1+v))^{\beta_1-1} (u/(1+v))^{\beta_2-1} \frac{u}{(1+v)^2}. \qquad (3.28)$$

The PDF of u is obtained by integrating out v as

$$f(u) = \frac{\alpha^{\beta_1+\beta_2}}{\Gamma(\beta_1)\Gamma(\beta_2)} e^{-\alpha u} u^{\beta_1+\beta_2-1} \int_0^\infty v^{\beta_1-1}/(1+v)^{\beta_1+\beta_2} dv. \qquad (3.29)$$

Put $1/(1 + v) = t$ so that $v = (1 - t)/t$, and $dv = -1/t^2 dt$. This gives us

$$f(u) = \frac{\alpha^{\beta_1+\beta_2}}{\Gamma(\beta_1)\Gamma(\beta_2)} e^{-\alpha u} u^{\beta_1+\beta_2-1} \int_0^1 t^{\beta_2-1}(1-t)^{\beta_1-1} dt =$$

$$\frac{\alpha^{\beta_1+\beta_2}}{\Gamma(\beta_1)\Gamma(\beta_2)} e^{-\alpha u} u^{\beta_1+\beta_2-1} B(\beta_1, \beta_2). \qquad (3.30)$$

This simplifies to

$$f(u) = \frac{\alpha^{\beta_1+\beta_2}}{\Gamma(\beta_1 + \beta_2)} e^{-\alpha u} u^{\beta_1+\beta_2-1}, \qquad (3.31)$$

which is GAMMA(β_1, β_2).

The PDF of v is found by integrating out u as

$$f(v) = \frac{\alpha^{\beta_1+\beta_2}}{\Gamma(\beta_1)\Gamma(\beta_2)} \frac{v^{\beta_1-1}}{(1+v)^{\beta_1+\beta_2}} \int_{u=0}^\infty u^{\beta_1+\beta_2-1} e^{-\alpha u} du. \qquad (3.32)$$

This simplifies to f(v) = $\frac{\Gamma(\beta_1+\beta_2)}{\Gamma(\beta_1)\Gamma(\beta_2)} \frac{v^{\beta_1-1}}{(1+v)^{\beta_1+\beta_2}}$, which is BETA2($\beta_1, \beta_2$) (also called Pearson type VI distribution).

Ratio of pairwise independent distributions

Let X_i's be IID EXP($\lambda\theta$) for i = 1, 2, . . . ,m. Let Y_j's be IID EXP(θ) for j = 1, 2, . . . ,n. If X_i's and Y_j's are pair-wise independent, find the distribution of W = U/V = $\sum_{i=1}^m X_i / \sum_{j=1}^n Y_j$.

Solution As X_i's are IID, the joint PDF is the product of individual PDFs. We first use the MGF technique to find the distribution of numerator and denominator. The MGF of EXP($\lambda\theta$) is $M_x(t) = 1/[1 - \lambda\theta t]$. As the X_i's are IID, $M_u(t) = 1/[1 - \lambda\theta t]^m$. Similarly, $M_v(t) = 1/[1 - \theta t]^n$. These are the MGFs of gamma distributions. Hence W is the ratio of two independent gamma variates, whose distribution is found in Example 3.3.3.

From independent gamma to beta distribution

If X and Y are IID GAMMA(α, β_i), prove that X/(X + Y) is BETA-I distributed.

Solution We find the distribution of U = X + Y and V = X/(X + Y). The joint PDF is

$$f(x, y) = \frac{\alpha^{\beta_1+\beta_2}}{\Gamma(\beta_1)\Gamma(\beta_2)]} x^{\beta_1-1} y^{\beta_2-1} e^{-\alpha(x+y)}. \tag{3.33}$$

The inverse mapping is x = uv, y = u(1 − v), so that the Jacobian is 1-u. The joint PDF of u and v is

$$f(u, v) = \frac{\alpha^{\beta_1+\beta_2}}{\Gamma(\beta_1)\Gamma(\beta_2)} (uv)^{\beta_1-1}(u - uv)^{\beta_2-1} e^{-\alpha u} u, \tag{3.34}$$

$0 < u < 1, 0 < v < \infty$. Combining common terms this becomes

$$\frac{\alpha^{\beta_1+\beta_2}}{\Gamma(\beta_1)\Gamma(\beta_2)} e^{-\alpha u} u^{\beta_1+\beta_2-1} v^{\beta_1-1}(1 - v)^{\beta_2-1}. \tag{3.35}$$

Integrating out u, it is easy to show that v has a BETA1(β_1, β_2) distribution.

Ratio of independent normal distributions

If x, y are IID N(0, σ_i^2), find the distribution of U = X/Y, and V = XY/$\sqrt{X^2 + Y^2}$.

Solution Here both U and V have range $-\infty$ to ∞. From U = X/Y, we get x = uy. Substituting for x in V, we get V = (uy)y/$\sqrt{(uy)^2 + y^2}$. Taking the y^2 within the square-root in the denominator outside, and canceling it out with the y in the numerator, this becomes $v = uy/\sqrt{u^2 + 1}$ so that $y = v\sqrt{u^2 + 1}/u$, and $x = v\sqrt{u^2 + 1}$. The Jacobian is easily obtained as

$$J = \begin{vmatrix} \frac{u}{v}/\sqrt{u^2+1} & \sqrt{u^2+1} \\ \frac{-v}{u^2\sqrt{u^2+1}} & \sqrt{u^2+1}/u \end{vmatrix} = v(1 + 1/u^2).$$

We first find the distribution of U. The joint PDF of X and Y is

$$f(x, y) = \frac{1}{2\pi\sigma_1\sigma_2} e^{-\frac{1}{2}(x^2/\sigma_1^2 + y^2/\sigma_2^2)}. \tag{3.36}$$

Substituting the values of x and y, $(x^2/\sigma_1^2 + y^2/\sigma_2^2) = v^2(1 + u^2)\left(\frac{1}{\sigma_1^2} + \frac{1}{\sigma_2^2 u^2}\right)$. Write A $= (1 + u^2)(\frac{1}{\sigma_1^2} + \frac{1}{\sigma_2^2 u^2})$, which is independent of v. Then f(u, v) = $\frac{1}{2\pi\sigma_1\sigma_2} e^{-\frac{1}{2}Av^2}$. Multiply by the Jacobian, and integrate out v, to get the PDF of u (we have to multiply by 2 due to the symmetry of the region of integration) as

$$f(u) = \frac{1}{\pi\sigma_1\sigma_2}(1 + 1/u^2) \int_{-\infty}^{\infty} v e^{-\frac{1}{2}Av^2} dv. \tag{3.37}$$

Put $v^2/2 = t$ in (3.37), so that $vdv = dt$, to get

$$f(u) = \frac{1}{\pi \sigma_1 \sigma_2}(1 + 1/u^2) \int_0^\infty e^{-At}\,dt = \frac{1}{\pi \sigma_1 \sigma_2}(1 + 1/u^2)[0 + 1/A]. \tag{3.38}$$

Upon substituting $A = (1 + u^2)(\frac{1}{\sigma_1^2} + \frac{1}{\sigma_2^2 u^2})$ this becomes

$$f(u) = \frac{1}{\pi \sigma_1 \sigma_2}(1 + 1/u^2)\frac{1}{(1 + u^2)}[1/(\frac{1}{\sigma_1^2} + \frac{1}{\sigma_2^2 u^2})]. \tag{3.39}$$

As $U = X/Y$ takes the value $k < 0$ when $(x > 0$ and $y < 0)$ or $(x < 0$ and $y > 0)$ and vice versa for $k > 0$, we have to multiply by 2 to get the correct density

$$f(u, \sigma_1, \sigma_2) = \frac{\sigma_1 \sigma_2}{\pi}\frac{1}{(\sigma_1^2 + u^2 \sigma_2^2)}, \quad -\infty < u < \infty. \tag{3.40}$$

This is the PDF of a scaled Cauchy distribution with scaling factor σ_2/σ_1. When $\sigma_1 = \sigma_2$, (3.40) reduces to the standard Cauchy distribution. Thus the ratio of two independent normals (with same variance) is Cauchy distributed. This result can be used to characterize the normal law as follows:

Remark 3.1 If X and Y are independent random variables with the same variance, whose ratio is Cauchy distributed, then X and Y are $N(0, \sigma^2)$ distributed.

We have assumed the independence of the normal variates in the above derivation. If they are dependent with correlation ρ, the ratio is no longer Cauchy distributed.

To find the PDF of V, integrate out u from $-\infty$ to ∞. Write the exponent as

$$v^2(1 + u^2)\left(\frac{1}{\sigma_1^2} + \frac{1}{\sigma_2^2 u^2}\right) = v^2\left[(1/\sigma_1^2 + 1/\sigma_2^2) + (u^2/\sigma_1^2 + 1/(\sigma_2^2 u^2))\right]. \tag{3.41}$$

Multiply by the Jacobian, and integrate out u, to get the PDF of v as

$$f(v) = \frac{1}{2\pi \sigma_1 \sigma_2}ve^{-\frac{1}{2}Av^2}\int_{-\infty}^\infty e^{-\frac{1}{2}v^2\left(\frac{u^2}{\sigma_1^2} + \frac{1}{\sigma_2^2 u^2}\right)}(1 + 1/u^2)\,du, \tag{3.42}$$

where A = $(1/\sigma_1^2 + 1/\sigma_2^2)$. Split the integral

$$\int_{-\infty}^{\infty} e^{-\frac{1}{2}v^2\left(\frac{u^2}{\sigma_1^2}+\frac{1}{\sigma_2^2 u^2}\right)}(1+1/u^2)du = \int_{-\infty}^{\infty} e^{-\frac{1}{2}v^2\left(\frac{u^2}{\sigma_1^2}+\frac{1}{\sigma_2^2 u^2}\right)}du+$$

$$\int_{-\infty}^{\infty} e^{-\frac{1}{2}v^2\left(\frac{u^2}{\sigma_1^2}+\frac{1}{\sigma_2^2 u^2}\right)}(1/u^2)du = I_1 + I_2 \text{(say)} \tag{3.43}$$

To evaluate I_1, we use the formula

$$\int_{-\infty}^{\infty} e^{-ax^2-b/x^2}dx = \sqrt{\frac{\pi}{a}}e^{-2\sqrt{ab}} \tag{3.44}$$

where a $= v^2/(2\sigma_1^2)$, and b $= v^2/(2\sigma_2^2)$, so that $\sqrt{ab} = v^2/(2\sigma_1\sigma_2)$. Thus $I_1 = \frac{\sigma_1}{v}\sqrt{2\pi}e^{-\frac{v^2}{\sigma_1\sigma_2}}$. To evaluate I_2, we make a simple substitution t = 1/u, so that dt $= -du/u^2$. Exponent of the integrand is $\left(\frac{u^2}{\sigma_1^2} + \frac{1}{\sigma_2^2 u^2}\right) = \left(\frac{1}{\sigma_1^2 t^2} + \frac{t^2}{\sigma_2^2}\right)$. Hence $I_2 = I_1$ (with σ_1 and σ_2 swapped) = $\frac{\sigma_2}{v}\sqrt{2\pi}e^{-\frac{v^2}{\sigma_1\sigma_2}}$. Substitute these values in (3.44) to get the PDF of v as

$$f(v) = \frac{1}{2\pi\sigma_1\sigma_2}ve^{-\frac{1}{2}Av^2}\left[\frac{\sigma_1}{v}\sqrt{2\pi}\,e^{-\frac{v^2}{\sigma_1\sigma_2}} + \frac{\sigma_2}{v}\sqrt{2\pi}\,e^{-\frac{v^2}{\sigma_1\sigma_2}}\right]. \tag{3.45}$$

Canceling out common factors (v, $\sqrt{2\pi}$ from numerator and denominator) and taking the $\sigma_1\sigma_2$ in the denominator into the brackets, we simplify this to the form

$$\frac{1}{\sqrt{2\pi}}e^{-\frac{v^2}{2}\left[A+\frac{2}{\sigma_1\sigma_2}\right]}(1/\sigma_1 + 1/\sigma_2). \tag{3.46}$$

Substitute A = $(1/\sigma_1^2 + 1/\sigma_2^2)$, and note that $(1/\sigma_1^2 + 1/\sigma_2^2 + \frac{2}{\sigma_1\sigma_2}) = (1/\sigma_1 + 1/\sigma_2)^2$, we find that this is the PDF of a normal distribution with variance $(\sigma_1\sigma_2/[\sigma_1 + \sigma_2])^2$.

Remark 3.2 If X and Y are independent continuous random variables with variances σ_1^2 and σ_2^2 such that $XY/\sqrt{X^2 + Y^2}$ is $N(0,(\sigma_1\sigma_2/[\sigma_1 + \sigma_2])^2)$ distributed, then X and Y are $N(0, \sigma_i^2)$ distributed.

Functions of independent chi-square

If X~ χ_m^2, Y~ χ_n^2 and Z~ χ_p^2 are independent, find the distribution of U = X/Y, V = (X + Y)/Z and W = X + Y + Z.

Solution Cross-multiply to get X = UY, (X + Y) = VZ, W = (X + Y) + Z = VZ + Z = (V + 1)Z. Eliminate Y between the first two relations and substitute for Z = W/(V + 1) to get Y = VW/[(U + 1)(V + 1)] and X = UVW/[(U + 1)(V + 1)]. As the ratio of two independent χ^2

variates are F distributed, U has an unscaled F (which is BETA-II) distribution. As X + Y is χ^2_{m+n}, V is BETA $-$ II(m + n, p) and W is χ^2_{m+n+p} distributed.

Function of gamma distribution

If X and Y are IID Gamma distributed with parameters (p, m) and (q, m) find the distribution of X/Y and (X + Y).

Solution From this we get $x = \frac{uv}{1+v}$, and $y = \frac{u}{1+v}$. The Jacobian of the transformation is

$$|J| = \begin{vmatrix} \frac{v}{1+v} & \frac{u}{(1+v)^2} \\ \frac{1}{1+v} & -\frac{u}{(1+v)^2} \end{vmatrix} = \frac{-u}{(1+v)^2}.$$

The joint PDF of X and Y is $f(x, y) = \frac{\alpha^{\beta_1+\beta_2}}{\Gamma(\beta_1)\Gamma(\beta_2)}e^{-\alpha(x+y)}x^{\beta_1-1}y^{\beta_2-1}$. Multiply by the Jacobian, and substitute for x, y to get

$$f(u, v) = \frac{\alpha^{\beta_1+\beta_2}}{\Gamma(\beta_1)\Gamma(\beta_2)}e^{-\alpha u}(uv/(1+v))^{\beta_1-1}(u/(1+v))^{\beta_2-1}\frac{u}{(1+v)^2}. \qquad (3.47)$$

The PDF of u is obtained by integrating out v as

$$f(u) = \frac{\alpha^{\beta_1+\beta_2}}{\Gamma(\beta_1)\Gamma(\beta_2)}e^{-\alpha u}u^{\beta_1+\beta_2-1}\int_0^\infty v^{\beta_1-1}/(1+v)^{\beta_1+\beta_2}dv. \qquad (3.48)$$

Put $1/(1 + v) = t$ so that $v = (1 - t)/t$, and $dv = -1/t^2 dt$. This gives us

$$f(u) = \frac{\alpha^{\beta_1+\beta_2}}{\Gamma(\beta_1)\Gamma(\beta_2)}e^{-\alpha u}u^{\beta_1+\beta_2-1}\int_0^1 t^{\beta_2-1}(1-t)^{\beta_1-1}dt =$$

$$\frac{\alpha^{\beta_1+\beta_2}}{\Gamma(\beta_1)\Gamma(\beta_2)}e^{-\alpha u}u^{\beta_1+\beta_2-1}B(\beta_1, \beta_2). \qquad (3.49)$$

This simplifies to

$$f(u) = \frac{\alpha^{\beta_1+\beta_2}}{\Gamma(\beta_1 + \beta_2)}e^{-\alpha u}u^{\beta_1+\beta_2-1}, \qquad (3.50)$$

which is GAMMA(β_1, β_2).

The PDF of v is found by integrating out u as

$$f(v) = \frac{\alpha^{\beta_1+\beta_2}}{\Gamma(\beta_1)\Gamma(\beta_2)}\frac{v^{\beta_1-1}}{(1+v)^{\beta_1+\beta_2}}\int_{u=0}^\infty u^{\beta_1+\beta_2-1}e^{-\alpha u}du. \qquad (3.51)$$

This simplifies to $f(v) = \frac{\Gamma(\beta_1+\beta_2)}{\Gamma(\beta_1)\Gamma(\beta_2)} \frac{v^{\beta_1-1}}{(1+v)^{\beta_1+\beta_2}}$, which is BETA2($\beta_1, \beta_2$) (also called Pearson type VI distribution). From these we can find the distribution of X/(X + Y) easily.

Bivariate beta

If X_1, X_2, X_3 are IID GAMMA(m_k,p) for k = 1,2,3 prove that the joint distribution of $Y_1 = X_1/X_3$ and $Y_2 = X_2/X_3$ is bivariate BETA-II(m_k) with PDF $f(y_1, y_2) = \frac{\Gamma(m_1+m_2+m_3)}{\Gamma(m_1)\Gamma(m_2)\Gamma(m_3)}$ $\frac{y_1^{m_1-1} y_2^{m_2-1}}{(1+y_1+y_2)^{m_1+m_2+m_3}}$, $y_1, y_2 > 0$.

Solution We know that the ratio of two independent GAMMA(m_k, p) is BETA-II distributed. As $Y_1 = X_1/X_3$ and $Y_2 = X_2/X_3$ are both BETA-II, their joint distribution is bivariate beta-II.

3.4   Polar Transformations

Polar transformation finds applications in integral calculus, differential equations, statistics, image-based computing (log-polar transformations) among many other fields. It is so called because the Cartesian points that use horizontal and vertical coordinate axes are transformed into polar coordinates that use radius and angle wrt a fixed set of coordinate axes.

3.4.1   Plane-Polar Transformations (PPT)

The name comes from the fact that it is applied in 2D for Cartesian to polar mapping. Let (x, y) represent the Cartesian coordinates and (r, θ) denote the corresponding polar coordinates. Then the mapping is defined by the relation x = r cos(θ) and y = r sin(θ). The inverse relation is r = $\sqrt{x^2 + y^2}$, $\theta = \tan^{-1}(y/x)$. Jacobian of the transformation is

$$J = \begin{vmatrix} \frac{\partial x}{\partial r} & \frac{\partial x}{\partial \theta} \\ \frac{\partial y}{\partial r} & \frac{\partial y}{\partial \theta} \end{vmatrix} = \begin{vmatrix} \cos(\theta) & -r\sin(\theta) \\ \sin(\theta) & r\cos(\theta) \end{vmatrix} = r.$$

As $x^2 + y^2 = r^2$, this transformation is especially useful in statistics when the PDF contains functions of the form x^2 or $1 \pm x^2$ (in the univariate case) and $\sum_i x_i^2$ or $1 \pm \sum_i x_i^2$ (in the multivariate case).

Find distribution of $\sqrt{X^2 + Y^2}$ of IID normal N(0, σ^2)

If X and Y are IID normal N(0, σ^2), find the PDF of $\sqrt{X^2 + Y^2}$.

Solution As they are independent, the joint PDF is

$$f(x, y) = \frac{1}{\sigma\sqrt{2\pi}}e^{-x^2/2\sigma^2}\frac{1}{\sigma\sqrt{2\pi}}e^{-y^2/2\sigma^2} = \frac{1}{2\pi\sigma^2}e^{-(x^2+y^2)/2\sigma^2}. \tag{3.52}$$

Make the transformation $x = r\cos(\theta)$, and $y = r\sin(\theta)$, so that $x^2 + y^2 = r^2$, and $\theta = \tan^{-1}(y/x)$. The Jacobian is r. Hence the joint PDF of r and θ is $f(r,\theta) = \frac{1}{2\pi\sigma^2}re^{-r^2/2\sigma^2}$. The density of r is found by integrating out θ. As $(x = +k, y = -k)$ and $(x = -k, y = +k)$ both map to the same value of r, we need to multiply the resulting density by 2 to get the correct PDF as

$$f(r, \sigma) = \int_{\theta=-\pi/2}^{\pi/2} \frac{1}{2\pi\sigma^2}re^{-r^2/2\sigma^2}d\theta = \frac{1}{\sigma^2}re^{-r^2/2\sigma^2}, 0 \leq r < \infty. \tag{3.53}$$

This is the Rayleigh distribution. As the joint PDF of r and θ is independent of θ, this is an indication that θ is uniformly distributed.

Marginal distribution of r and θ

Consider a circle with radius R centered at the origin. Let (x, y) be any point within the circle $(x^2 + y^2 \leq R^2)$. If r is the radial distance from the center $(r = \sqrt{x^2 + y^2}$ and θ is the angle made with the positive X-axis, find the marginal distribution of r and θ.

Solution As the total area is unity, the joint PDF of x and y is $f(x, y) = 1/\pi R^2, x^2 + y^2 \leq r^2$. The joint CDF of r and θ is $F(r,\theta) = 1/(\pi R^2)(\theta/(2\pi))\pi r^2 = r^2\theta/(2\pi R^2)$. Differentiate wrt r and θ to get the PDF as $f(r,\theta) = r/\pi R^2$, $-R<r<R$, $0< \theta < 2\pi$. From this the marginal distribution of r is obtained as $f(r) = \int_0^{2\pi} f(r, \theta)d\theta = 2r/R^2$ for $-R<r<R$. Similarly $g(\theta) = \int_{r=0}^{R} f(r, \theta)dr = \int_{r=0}^{R} 2r/R^2 dr = R^2/(2\pi)$ for $0\leq \theta < 2\pi$.

From IID χ_n^2 to Student's T distribution

If X and Y are IID χ_n^2 distributions, prove that $Z = \frac{\sqrt{n}}{2}\frac{X-Y}{\sqrt{XY}}$ has a Student's T distribution.

Solution Consider the polar transformation $x = r\cos^2(\theta)$, and $y = r\sin^2(\theta)$. Then $x + y = r(\sin^2(\theta) + \cos^2(\theta)) = r$, $x - y = r(\cos^2(\theta) - \sin^2(\theta)) = r\cos(2\theta)$, $\sqrt{xy} = r\sin(\theta)\cos(\theta) = \frac{r}{2}\sin(2\theta)$, so that $xy = \frac{r^2}{4}\sin^2(2\theta)$. Jacobian of the transformation is

$$J = \begin{vmatrix} \frac{\partial x}{\partial r} & \frac{\partial x}{\partial \theta} \\ \frac{\partial y}{\partial r} & \frac{\partial y}{\partial \theta} \end{vmatrix} = \begin{vmatrix} \cos^2(\theta) & -r\sin(2\theta) \\ \sin^2(\theta) & r\sin(2\theta) \end{vmatrix} = r\sin(2\theta).$$

Substituting the values of X and Y in $Z = \frac{\sqrt{n}}{2}\frac{X-Y}{\sqrt{XY}}$, we get $Z = \frac{\sqrt{n}}{2}2\cot(2\theta) = \sqrt{n}\cot(2\theta)$. The joint PDF of X and Y is

$$f(x,y) = \frac{1}{2^n\Gamma(n/2)^2}e^{-(x+y)/2}(xy)^{\frac{n}{2}-1}. \tag{3.54}$$

Multiply by the Jacobian and substitute the values of x and y to get

$$f(r,\theta) = \frac{1}{2^n\Gamma(n/2)^2}e^{-r/2}\left(\frac{r^2}{4}\sin^2(2\theta)\right)^{\frac{n}{2}-1}r\sin(2\theta). \tag{3.55}$$

This after simplification reduces to

$$f(r,\theta) = \frac{r^{n-1}e^{-r/2}}{2^{2n-2}\Gamma(n/2)^2}\sin(2\theta)^{n-1}. \tag{3.56}$$

Distribution of θ is obtained by integrating out r. Thus

$$f(\theta) = \frac{\sin(2\theta)^{n-1}}{2^{2n-2}(\Gamma(n/2))^2}\int_{r=0}^{\infty}r^{n-1}e^{-r/2}dr =$$

$$\frac{\sin(2\theta)^{n-1}}{2^{2n-2}(\Gamma(n/2))^2}2^n\Gamma(n) = \frac{\Gamma(n)}{2^{n-2}(\Gamma(n/2))^2}\sin(2\theta)^{n-1}. \tag{3.57}$$

Consider the transformation $t = \sqrt{n}\cot(2\theta)$, so that $\frac{\partial\theta}{\partial t} = 1/[2\sqrt{n}\csc^2(2\theta)]$ $= 1/[2\sqrt{n}(1+\cot^2(2\theta)]$. Writing $\sin(2\theta) = 1/\sqrt{1+\cot^2(2\theta)}$, and multiplying by the Jacobian we get

$$f(t,n) = \frac{\Gamma(n)}{\sqrt{n}2^{n-1}\Gamma(n/2)^2}\frac{1}{(1+t^2/n)^{(n+1)/2}}. \tag{3.58}$$

Multiply the numerator and denominator by $\Gamma((n+1)/2)\Gamma(1/2)$ and use the formula $\Gamma(n)\Gamma(1/2) = 2^{n-1}\Gamma(n/2)\Gamma(\frac{n+1}{2})$ to get the constant multiplier in the form $1/[\sqrt{n}B(\frac{1}{2},\frac{n}{2})]$ (where $\Gamma(1/2) = \sqrt{\pi}$). This is the Student T distribution T(n). See [Cacoullos (1965)] for a CDF derivation of this and related results.

Functions of IID Weibull distributions

If X, Y are IID Weibull(2,b) with PDF $f(x,b) = \frac{2}{b^2}xe^{-x^2/b^2}$, for $x \geq 0$, show that $Z = XY/(X^2 + Y^2)$ has PDF $f(z) = 2z/\sqrt{1-4z^2}$, which is independent of b.

Solution Put $x = r\cos(\theta)$, and $y = r\sin(\theta)$, so that $x^2 + y^2 = r^2$, and $\theta = \tan^{-1}(y/x)$. The Jacobian of the transformation is r. Hence the joint PDF of r and θ is $f(r,\theta) = \frac{4r^3}{b^4}\sin(\theta)\cos(\theta)e^{-r^2/b^2}$. Using $2\sin(\theta)\cos(\theta) = \sin(2\theta)$ this becomes $f(r,\theta) = \frac{2r^3}{b^4}\sin(2\theta)$ e^{-r^2/b^2}. The PDF of θ is obtained by integrating out r as

$$f(\theta) = \frac{2\sin(2\theta)}{b^4} \int_0^\infty r^3 e^{-r^2/b^2} dr. \tag{3.59}$$

Put $r^2/b^2 = t$, so that $r\,dr = (b^2/2)dt$. Then $f(\theta) = \frac{2\sin(2\theta)}{b^4}\frac{b^4}{2}\int_0^\infty te^{-t}dt = \sin(2\theta)$. Putting the values of x and y in Z gives $Z = \frac{1}{2}\sin(2\theta)$ so that $\frac{\partial z}{\partial \theta} = \cos(2\theta) = \sqrt{1-4z^2}$. Hence f(z) $= 2z/\sqrt{1-4z^2}$, $-\frac{1}{2} \le z \le \frac{1}{2}$.

3.4.2   Cylindrical Polar Transformations (CPT)

This transformation is a simple extension of the above to 3D. The mapping is defined by the relations $x = r\cos(\theta)$, $y = r\sin(\theta)$, and $z = z$. Jacobian of this transformation also is r (see Table 3.2). This defines a cylinder of base r in the polar coordinates. Inverse mapping is easily obtained as $r = \sqrt{x^2 + y^2}$, $\theta = \tan^{-1}(y/x)$.

3.4.3   Spherical Polar Transformations (SPT)

This is a more general form of the above PPT to 3D. The mapping is defined by the relations $x = r\cos(\theta)\cos(\phi)$, $y = r\sin(\theta)\cos(\phi)$, and $z = r\sin(\phi)$, so that $x^2 + y^2 + z^2 = r^2\cos^2(\phi)[\cos^2(\theta) + \sin^2(\theta)] + r^2\sin^2(\phi) = r^2$. The inverse mapping is defined as $r = \sqrt{x^2 + y^2 + z^2}$, $\theta = \tan^{-1}(y/x)$, and $\phi = \sin^{-1}(z/r)$. Jacobian of this transformation is $r^2\cos^2(\phi)$ (see Table 3.2). An equivalent mapping is defined by the relations x $= r\cos(\theta)\sin(\phi)$, $y = r\sin(\theta)\sin(\phi)$, and $z = r\cos(\phi)$. The inverse mapping is given by $r = \sqrt{x^2 + y^2 + z^2}$, $\theta = \tan^{-1}(y/x)$, and $\phi = \cos^{-1}(z/\sqrt{x^2 + y^2 + z^2})$, or $\phi = \tan^{-1}(\sqrt{x^2 + y^2}/z)$. Jacobian of this transformation is $-r^2\sin(\phi)$ (so that dx dy dz $= r^2\sin(\phi)$

Table 3.2 Common polar transformation of three variables

Name	Transformation	Jacobian $\|J\|$
Cylindrical	$x = r\cos(\theta)$, $y = r\sin(\theta)$, and $z = z$	r
Spherical	$x = r\cos(\theta)\cos(\phi)$, $y - r\sin(\theta)\cos(\phi)$, $z = r\sin(\phi)$	$r^2\cos^2(\phi)$
Spherical	$x = r\cos(\theta)\sin(\phi)$, $y = r\sin(\theta)\sin(\phi)$, $z = r\cos(\phi)$	$-r^2\sin(\phi)$
Toroidal	$x = (r\cos(\theta) + R)\cos(\phi)$, $y = (r\cos(\theta) + R)\sin(\phi)$, $z = r\sin(\theta)$	$r\cos(\theta)+R$
Toroidal	$x = r*\cos(\theta)\cos(\phi)$, $y = Cr*\sin(\theta)$, $z = Dr*\sin(\phi)$	$r^2(m^2\cos^2(\phi)+$ $n^2\cos^2(\theta))/[C*D]$

The sign of $|J|$ is ignored in statistical applications. Inverse of the transformations appear in respective sections. Last row has $C = \sqrt{1 - m^2\sin^2(\phi)}$, $D = \sqrt{1 - n^2\sin^2(\theta)}$

dr $d\theta$ $d\phi$). The name *spherical transformation* comes from the fact that its domain is of the form $x^2 + y^2 + z^2$ or arithmetic functions of it.

3.5   Other Methods

The SPT can be generalized to n-dimensions in multiple ways. One simple way is to use the Helmert transformation $x_1 = r \cos(\theta_1)$, $x_2 = r \sin(\theta_1) \cos(\theta_2)$, $x_3 = r \sin(\theta_1) \sin(\theta_2)$ $\cos(\theta_3), \ldots, x_{n-1} = r \sin(\theta_1) \sin(\theta_2) \ldots \sin(\theta_{n-2}) \cos(\theta_{n-3})$, and $x_n = r \sin(\theta_1) \sin(\theta_2) \ldots \sin(\theta_{n-1})$. The Jacobian is given by $|J| = r^{n-1} \sin^{n-2}(\theta_1) \sin^{n-3}(\theta_2) \sin^{n-4}(\theta_3) \cdots \sin(\theta_{n-2})$. Squaring and adding each term, we get $r^2 = x_1^2 + x_2^2 + \cdots + x_n^2$.

Toroidal polar transformation (TPT) is an extension of SPT, defined as

$$x = (r \cos(\theta) + R) \cos(\phi), \; y = (r \cos(\theta) + R) \sin(\phi), \text{ and } z = r \sin(\theta), \quad (3.60)$$

so   that $x^2 + y^2 + z^2 = (r \cos(\theta) + R)^2 [\cos^2(\phi) + \sin^2(\phi)] + r^2 \sin^2(\theta) = r^2 + R^2 + 2rR \cos(\theta)$. The inverse mapping is defined as

$$r = \{[(x^2 + y^2)^{1/2} - R]^2 + z^2\}^{1/2}, \; \phi = \tan^{-1}(y/x), \text{ and } \theta = \sin^{-1}(z/r). \quad (3.61)$$

The Jacobian is $1/J =$

$$\begin{vmatrix} \cos(\theta) \cos(\phi) & \cos(\theta) \sin(\phi) & \sin(\theta) \\ -B * \sin(\phi) & B * \cos(\phi) & 0 \\ -r \cos(\phi) \sin(\theta) & -r \sin(\phi) \sin(\theta) & r \cos(\theta) \end{vmatrix}$$

where B = (r $\cos(\theta)$+R). To evaluate this determinant, take out B from second row, r from third row, multiply new first column by $\cos(\phi)$, new second column by $\sin(\phi)$, and add new second column to the first column ($C_1 = C_1 + C_2$). The (2,1)th element also becomes zero. Then expand the determinant along second row to get the Jacobian as B = (r $\cos(\theta)$ + R).

Another general transformation is given by x = r $\cos(\theta)\cos(\phi)$, y = r $\sin(\theta)$ $\sqrt{1 - m^2 \sin^2(\phi)}$, z = r $\sin(\phi)\sqrt{1 - n^2 \sin^2(\theta)}$, where $m^2 + n^2 = 1$. Squaring and adding gives us $x^2 + y^2 + z^2 = r^2$. The Jacobian in this case is $1/J =$

$$\begin{vmatrix} \cos(\theta) \cos(\phi) & -r \sin(\theta) \cos(\phi) & -r \cos(\theta) \sin(\phi) \\ \sin(\theta) \sqrt{1 - m^2 \sin^2(\phi)} & r \cos(\theta) \sqrt{1 - m^2 \sin^2(\phi)} & \frac{-m^2 r \sin(\theta) \sin(\phi) \cos(\phi)}{\sqrt{1 - m^2 \sin^2(\phi)}} \\ \sin(\phi) \sqrt{1 - n^2 \sin^2(\theta)} & \frac{-n^2 r \sin(\theta) \sin(\phi) \cos(\theta)}{\sqrt{1 - n^2 \sin^2(\theta)}} & r \cos(\phi) \sqrt{1 - n^2 \sin^2(\theta)} \end{vmatrix}$$

To evaluate above, take r as a common factor from second and third columns, multiply first column by $\sin(\theta)$, second column by $\cos(\theta)$ and apply $C_1 = C_1 + C_2$ (ie. add new second column to new first column). The first element at (1, 1) reduces to zero, so that the determinant becomes that of two 2×2 matrices. Using the relationship $m^2 + n^2 = 1$, this is easily seen to be $r^2 (m^2 \cos^2(\phi) + n^2 \cos^2(\theta))/[\sqrt{1 - m^2 \sin^2(\phi)}\sqrt{1 - n^2 \sin^2(\theta)}]$.

There are many other ways to find the distribution of transformed variables. One possibility is to use the characteristic function (if it is easily invertible) of the original variates. Let $U = g(x_1, x_2, \ldots, x_n)$ be the transformation required. If $\phi_Z(t) = E(e^{itg(x)}) = \int \int e^{itg(x)} f(x_1, x_2, \ldots, x_n) dx_1 .. dx_n$ is easy to evaluate, we could simply use inversion theorem to get the PDF of U.

3.6   Rosenblatt Transformation

Rosenblatt transformation is a general technique to transform a group of two or more non-normal RVs into another group of equivalent and independent standard RVs. It uses conditional probability to transform non-normal RVs to IID standard distributions such as normal, standard uniform, or standard exponential RVs. Such transformations are often helpful in generating independent samples, simplifying complex systems modeling, and in individual risk assessments of variates. If $(V_1, V_2, \ldots, V_m)$ is a random vector that are possibly correlated, Rosenblatt transform constructs a sequence of IID RVs $U_1 = F(V_1)$, $U_2 = F(V_2|V_1)$, $U_3 = F(V_3|V_1 V_2)$ etc using conditional decomposition. The resulting random vectors $(U_1, U_2, \ldots, U_m)$ are independent.

3.7   Copulas

Copulas are multivariate distribution functions with uniformly distributed marginals on the real line as arguments. It provides the dependence structure among univariate marginal CDFs and the multivariate CDF. The name comes from Latin where it means to 'couple', 'link', 'bond', or 'connect' multiple parts into a single whole. The C() notation is universally used to represent copulas with or without a subscript n or a 'n-' prefix (n being the number of variables or dimensionality). It was introduced by Abe Sklar (1959) although the idea can be traced back in the works of Udny Yule and others during 1900s.

Definition 3.3 A copula is a 'connecting' or 'joiner' function used to describe or model dependency among two or more continuous RVs that expresses a joint CDF as a function of several marginal CDFs.

The basic principle used in copulas is that the CDF (as well as the survival function SF) of any continuous RV is uniformly distributed (Sect. 2.2.5). A bivariate copula is a function from $[0, 1]^2$ to $[0, 1]$ as $\Pr[X \leq x, y \leq y] = C(x, y)$ where X and Y are U[0, 1] distributed.[2] The Sklar's theorem given below (for the bivariate case) makes it precise.

---

[2] It is known as an aggregator in some fields. The plural form is either copulae or copulas.

Theorem *Let F(x, y) be the CDF of a bivariate random vector (x, y) with marginal CDFs F(x) and F(y). Then there exists a copula C such that for all (x, y)∈ $\mathbb{R}^2$, F(x, y) = C(F(x), F(y)).*

This concept has been extended when some of the RVs are discrete. Thus a copula C captures the dependence among the marginal CDFs irrespective of their actual distributions.

As F(x) and F(y) are bounded and finite taking values in [0, 1] for every possible values of x, and y; the copula will be a finite and continuous function. However the exact functional form is not easy to identify for distributions other than those whose inverse CDF has closed form (the CDF of uniform and exponential distributions are easily invertible). They are denoted mathematically for p variates as:

$$\mathrm{F}(x_1, x_2, \ldots, x_p) = C(F(x_1), F(x_2), \ldots, F(x_p)) = C(u_1, u_2, \ldots, u_p) \tag{3.62}$$

where $F(x_1, x_2, \ldots, x_p)$ is the CDF of a p-dimensional distribution and C() denotes the copula which is regarded as a 'connector' or 'joiner' of continuous function with p arguments (parameters) and returns the joint distribution in terms of marginal CDFs. Each of the terms on the RHS (being marginal CDFs) takes values in (0, 1). They may be permuted in any desired order, and need not belong to the same family. As examples, the distribution of BMI differs a lot in various age groups and localities, and the distribution of personal incomes are highly positively skewed whereas family incomes are less skewed due to the possibility of unemployed spouse in high income person's family. Thus we could form the copula of BMI from different localities and age-groups; or personal income of people in different education levels and combine them into a joint distribution. Similarly, the distribution of time-series data and error terms follow different laws but they could be combined using a time-invariant copula [Cherubini and Gobbi (2013)]. Whatever may be the marginal distribution of the RVs, they can be combined to get the joint distribution using a copula. The RHS is the probability that each of the RVs $(X_1, X_2, \ldots, X_p)$ assumes values less than $(x_1, x_2, \ldots, x_p)$. Replace each of the marginal CDFs on the right by independent unit-rectangular variates to get the general definition of a copula as $C(u) = \Pr[U_1 \leq u_1, U_2 \leq u_2, \ldots, U_p \leq u_p]$ where $U_j \sim U(0,1)$ RV. When both the variables are continuous, the bivariate copula can be considered geometrically as the C-volume of the rectangle [0, u]*[0, v] where (u, v) ∈ [0,1]. A survival copula results when the marginal CDFs on the RHS are replaced by corresponding survival functions (SFs) as $S(x_1, x_2, \ldots, x_p) =$

$$C(SF(x_1), SF(x_2), \ldots, SF(x_p)) = C(1 - u_1, 1 - u_2, \ldots, 1 - u_p) \tag{3.63}$$

where $SF(x_k) = 1 - F(x_k)$ is the marginal survival function (or survival probability in discrete case). Particular cases are $SF(x_1) = 1 - F(x_1)$, $SF(x_1, x_2) = 1 - F_1(x_1) - F_2(x_2) + F_{12}(x_1, x_2)$, $SF(x_1, x_2, x_3) = 1 - F_1(x_1) - F_2(x_2) - F_3(x_3) + F_{12}(x_1, x_2) + F_{13}(x_1, x_3) + F_{23}(x_2, x_3) - F_{123}(x_1, x_2, x_3)$, and so on. A symmetric copula ensues when it is permutation invariant. A copula is radially symmetric if $S(x_1, x_2, \ldots, x_p) - C(x_1, x_2, \ldots, x_p)$. Examples of radially symmetric copulae include that of n IID U(0, 1) variates, and Plackett copula that arises in the study of 2×2 contingency tables. All copula functions need not be differentiable. But when they are differentiable, we could express the joint PDF in terms of marginal PDFs.

The bivariate copulas (number of variables p = 2) is used in bi-objective feature selection problems encountered in binary classification. Then $F(x_1, x_2) = \Pr(X_1 \leq x_1, X_2 \leq x_2) = \Pr(U_1 \leq x_1, U_2 \leq x_2)$. As an example, the bivariate Gaussian copula is defined as

$$C(u, v) = \Phi_2(\Phi^{-1}(u), \Phi^{-1}(v)) \tag{3.64}$$

where $\Phi^{-1}(u)$ is the inverse CDF of N(0,1) and $\Phi_2()$ is the CDF of bivariate unit normal distribution. An example of a bivariate survival copula is SC(u, v) = u + v − 1 − C(1 − u, 1 − v).

An important property of copulas is that they are invariant under monotonically increasing transformations of the marginals. A direct implication of this result is that Spearman's ρ between two RVs X and Y is the same as Pearson's r between the marginals F(x) and G(y). Copulas can work in the reverse direction as well. If the univariate marginals of two or more RVs are known, a multivariate distribution exists with joint CDF $C(F(x_1), F(x_2), \ldots, F(x_p))$ whose marginals are $F(x_k)$ for k = 1, 2, ...,p. The marginals may be RVs of the same family in some applications. For instance, the crossover counts in various parts of a genome (in units of space or time) can be modeled using Poisson distributions with different parameters in genomics. Copulas can be considered as "joint distribution generating functions" that link marginal CDFs or SFs of RVs to their joint distributions. The converse states that any multivariate CDF has an associated copula (this follows directly from the above when each RV is continuous, in which case the copula is unique).

A copula can be indexed by additional parameters. As an example, the Ali-Mikhail-Haq (AMH) survival copula is given by $C_\theta(u, v) = uv/[1 − \theta(1 − u)(1 − v)]$ for $\theta \in [−1, +1)$. Copulas find applications in economics, banking and finance, investment and stock markets to model economic risks such as financial losses or deprecation of assets. It is used for portfolio risk assessment and hedge fund management. Hoeffding-Fréchet bounds on copulas is used to find the lower and upper bounds on risk factors that depend on random phenomena.

Vine copulas are graphical visualization models that use a cascade of bivariate copulas. They have several subtypes such as R-vines (regular), C-vines and D-vines. All constraints are two-dimensional in R-vines and are drawn as a top-down connected tree (with root at the bottom) [Allen et al. (2013)]. A canonical vine (C-vine) is a tree in which each subtree has a unique node of degree n − 1 (has the maximum degree). All nodes in a D-vine have degrees ≤ 2. They are used for risk-modeling in financial economics, actuarial and environmental sciences, etc.

3.7.1   Properties of Copulas

Copulas are multivariate functions whose arguments being marginal CDFs are always real numbers in [0,1]. As it represents the joint probability, the value returned by a copula is also in [0,1].

1. $C(0,0,\cdot,0) = 0$; $C(1, 1, 1, \ldots, u_k, 1, \ldots, 1) = u_k \forall k$.
2. $C(u_1, u_2, \ldots, u_p) = 0$ whenever at least one of the $u_k = 0$.
3. $C(1, 1, \ldots, u, \ldots, 1) = u$ whenever at most one of the $u_k = u$ and all others are 1.
4. If x and y are vectors of compatible dimensions then $|C(x) - C(y)| \le ||x - y||$ and $|C(u_2, v_2) - C(u_1, v_1)| \le |u_2 - u_1| + |v_2 - v_1|$ where $||x - y||$ is the L_1-norm.
5. $C(u) = H(F_1^{-1}(u_1), F_2^{-1}(u_2), \ldots, F_p^{-1}(u_p))$ where $F_k^{-1}(u_k)$ is the quasi-inverse (left-inverse)[3] of F defined as $F_k^{-1}(u) := \inf(x: F_k(x) \ge u)$.
6. For every $u_1, u_2, v_1, v_2 \in [0,1]$, if $u_1 \le u_2$ and $v_1 \le v_2$ then $C(u_2, v_2) - C(u_2, v_1) - C(u_1, v_2) + C(u_1, v_1) \ge 0$.
7. $C()$ is non-decreasing in each of the p variables.
8. $C(u_1, u_2, \ldots, u_p) = F(F^{-1}(u_1), F^{-1}(u_2), \ldots, F^{-1}(u_p))$.
9. If X and Y are continuous, they are independently distributed iff $C(u, v) = uv$. In general, if the copula is the product of its arguments, it means that the variables are independent (so that the joint CDF is the product of marginal CDFs).[4]
10. If X and Y are related through a continuous increasing function Y = g(X), then $C(u, v) = \min(u, v)$.
11. If X and Y are related through a continuous decreasing function Y = g(X), then $C(u, v) = \max(u + v - 1, 0)$.
12. Any convex combination of continuous copulas is itself a copula.

The item 5 above tells us that an implicit expression for the copula function can be obtained when the joint PDF and inverse of marginal PDFs are available. However, the joint PDF is usually unavailable and is either obtained from prior studies or approximated from domain knowledge as Gaussian, log-normal, skew-normal, Pareto or extreme value distribution. The risk profile of portfolios in banking and stock markets is assumed to be log-normal or skew-normal distributed, which could only be approximate as it is time-dependent. In addition, several asset returns are fat-tailed for which Pareto or extreme value distributions are more popular.[5] As copulae are invariant under strictly increasing (monotonic) transformations, any closure property of the joint CDF which is invariant under strictly increasing

---

[3] The left-inverse or left-continuous-inverse of a function is defined as $F^{-}(u) = \inf\{x \in \mathbb{R} : F(x) \ge u\}$ for $0 < u < 1$.

[4] independent copulas are usually denoted by Π as $\Pi(x_1, x_2, \ldots, x_n) = x_1 * x_2 * \cdots x_n$.

[5] Another choice is the 3-parameter Laplace distribution with PDF $f(x; \mu, \sigma, p) = K \exp(-\frac{1}{p}|\frac{x-\mu}{\sigma_p}|^p)$ where $1/K = 2\ p^{1/p}\Gamma(1 + 1/p)\sigma_p$, which reduces to the Laplace distribution for $p = \sigma_p = 1$ and normal distribution for $p = 2$.

transformations is a property of its copula. They may also be defined for order statistics. If $X = (x_1, x_2, \ldots, x_n)$ and $Y = (y_1, y_2, \ldots, y_n)$ are two independent samples from two populations with marginal CDFs F1 and F2, the copula of $x_{(n)}$ and $y_{(n)}$, the respective maximums is called extreme value (EV) copula. The EV copula has an interesting property that $C(u, v) = C(u^{1/n}, v^{1/n})^n \forall u, v \in [0,1]$. They find applications in modeling extreme events in hydrology, geology, and natural disasters such as earthquakes, blizzards, droughts, etc. The Archimedian copula is defined as $C(u, v) = f(f^{-1}(u) + f^{-1}(v))$ where f() is an invertible function [Groesser and Okhrin (2021)]. From this the k-D multivariate Archimedian copula follows as $C(u_1, u_2, \ldots, u_k) = f(\sum_k f^{-1}(u_k))$, where $u_1, u_2, \ldots, u_k \in [0, 1]$ with f(0) = 1 and $f(\infty) = 0$ (which is called generator which must be differentiable up to order k-2 with $(-1)^j f^{(j)}(x) \geq 0$ [McNeil and Neslehova (2008)]. For RVs defined on the unit interval, we could define a special copula as $C(u, v) = F(F_1^{-1}(u), F_2^{-1}(v))$ where $F(F_k^{-1}())$ is the inverse CDF (Chap. 2).

Copulas can be used to express conditional density functions. The bivariate conditional copula is

$$f(x_j|x_i) = f(x_i, x_j)/f_i(x_i) = c_{ij}(F_i(x_i), F_j(x_j)) \cdot f_j(x_j) \tag{3.65}$$

which for i = 1 and j = 2 reduces to

$$f(x_2|x_1) = f(x_1, x_2)/f_1(x_1) = c_{12}(F_1(x_1), F_2(x_2)) \cdot f_2(x_2). \tag{3.66}$$

Similarly,

$$f(x_3|x_1, x_2) = c_{13|2}(F_{1|2}(x_1|x_2), F_{3|2}(x_3|x_2)) \cdot f_{3|2}(x_3|x_2). \tag{3.67}$$

If the marginal of V is uniform, $F(u|v) = (\partial/\partial v) C(u, v) = c(u, v)$ and vice-versa for U.

It is well-known that Pearson's correlation is invariant under strictly linear transformations (of either or both variables), but not under monotonic transformations; whereas Spearman's, Kendall's and Chatterjee's correlations are invariant under monotonic transformations [Chattamvelli (2024)]. Rank-based empirical copulas can also be developed using indicator functions as

$$C_n(u, v) = \frac{1}{n} \sum_{i=1}^{n} I(R_i/(n+1) \leq u, S_i/(n+1) \leq v), \tag{3.68}$$

where I() is the indicator function. Kendall's correlation can be expressed in terms of copulas as

$$\tau(x, y) = 4 \int_{[0,1]} \int_{[0,1]} C(u, v) \, dC(u, v) - 1 \tag{3.69}$$

and Spearman's correlation can be expressed as

$$\rho(x, y) = 12 \int_{[0,1]} \int_{[0,1]} uv \, dC(u, v) - 3 = \frac{n-1}{n+1} \rho_n \tag{3.70}$$

where $dC(u, v) = (\partial^2/\partial u \partial v) C(u, v)$.

Convolutional copulas are obtained by the convolution operator and finds applications in econometrics [Cherubini and Gobbi (2013)]. Copula transformation is a technique to impute multiple missing values by transforming any continuous multivariate non-normal data to multivariate normal [Lun and Khattree (2024)]. A copula being a multivariate combiner is a continuous function. It can be differentiated wrt the variables to get a copula density (which is denoted by lowercase c()) as $c(\mathbf{u}) = \partial^p/[\partial u_1 \partial u_2 \cdots \partial u_p]C(\mathbf{u})$. This can be found from joint and marginal PDFs as $c(\mathbf{u}) = f(x)/[f_1(x_1)f_2(x_2) \cdots f_p(x_p)]$ where $x_j = F_j^{-1}(u_j)$.

3.7.2 Copula Dependent Measures

There are several dependent measures that can be obtained from copulas. Consider $\lambda_z = \lim_{z \to 1^-} S(z,z)/(1-z)$ where z is either u or v, and the limit is from below. This results in two dependent measures λ_u and λ_v. λ_u is the conditional probability that $\Pr[U > u|V > u]$ and similarly $\lambda_L = \Pr[U < v|V < v]$.

3.8   Applications

Product of IID random variables has many applications. As an example, the product of triangular distributions is used in oil-exploration and nuclear waste management [Glickman and Xu (2008)]. Distribution of products of correlated zero-mean Gaussian RVs can be expressed in terms of Bessel functions [Gaunt (2022)]. Highly accurate approximations to sums, products and ratios of two RVs can also be found using a linear combination of exponentials [Beylkin et al. (2019)]. It is also used in various processes connected in series. For instance, if one technique magnifies an image by p% and another technique works on the magnified image (as input) and further magnifies by q%, the total magnification achieved is the product of p and q. Other similar processes are sequential purifications (say of ores), magnification of signals (such as radio signals or gravitational waves from far deep-space), application of sequential force multipliers (such as compression pressures), etc. It is also used in actuarial sciences and economics to model risks which are dependent on additive economics variables (inflation, interest rates, GDP growth).

Ratio of RVs is used in inventory control (inventory ratios of fast moving items), meteorology (precipitation ratios), and medical sciences (in modeling ratio measures such as cardiac index mentioned in Chap. 1). Similarly, if X denotes the fuel consumption of a vehicle and Y denotes the total payload, the distribution of the ratio X/Y for various vehicles of the same type is of interest to transportation engineers. As shown in Sect. 3.3, the difference of IID exponential distributions is Laplace distributed. It is used in data communications and transportation engineering. Very large files are sometimes downloaded from multiple servers through distinct channels due to communication bandwidth limitations or router bottlenecks. Suppose the waiting time for each such channel is exponentially distributed with the same parameter. Then the difference in waiting times is Laplace distributed. The position errors of aircraft or sea-vessels in motion along predetermined paths are usually obtained by pooling data from complex navigation systems such as GPS. The distribution of these errors in fixed time intervals is modeled using the Laplace distribution. If the lateral currents follow an exponential distribution, we have a convolution of exponential and Laplace distributions for the resultant speed.[6]

Another approach to find the distributions of X/Y, X/(X + Y) etc. using copulas is described in [Ly et al. (2019b)]. The result in Remark 3.2 on the distribution of U = $2XY/\sqrt{X^2 + Y^2}$ is applicable to all spherically symmetric distributions [Jones (1999)]. Breiman's theorem is a simple method to find the distribution of products of dependent random variables [Chen et al. (2019); Cadenza et al. (2022)]. Dirac's δ-method can be used to find the distribution of simple transformations [Au and Tam (1999)]. See [Annavajjala et al. (2010)] for the distribution of ratios of exponential variates, and [Bailey (1992)] for ratios of gamma variates.

The total seismic moment used in seismology is the sum of the moments of individual seismic events at a specific location. If each event is represented as a RV, this is the sum of IID RVs. Heavy tailed distributions such as the Pareto law gives good fit for hazard assessment. Fading channels in wireless communications are often modeled as random processes, and the signal strength at a particular time is a RV. Applications of the joint distribution of the sum and maximum of exponential laws appear in [Arendarczyk et al. (2018b)], and dependent Pareto risks in [Arendarczyk et al. (2018a)].

3.9   Summary

This chapter discusses the methodology to obtain the marginal, joint, and conditional probability distributions for both the discrete and continuous distributions. The concept and tools for the Jacobian to derive the joint probability distribution of functions of continuous random variables are introduced and illustrated in this chapter. Distribution of functions of 2 or more variates has received much attention in the literature. Most of the research in this field uses the normal ([Quine (1994)]) and exponential ([Annavajjala et al. (2010)]) distributions

See [Shepp (1964); Quine (1994); Baringhaus et al. (1988)], and [Jones (1999)] for alternative derivations of the result in Example 3.3.3. [Bansal et al. (1999)] uses the uniqueness of moments to prove that the distribution of $2XY/\sqrt{X^2 + Y^2}$ is identical to that of X and Y. This chapter also includes a brief discussion on copulas. An important property of copulas is that it is invariant under strictly monotone transformations.

---

[6] Lateral currents quite often have approximately uniform (rectangular) distribution, but exponential distribution is a better fit for strong turbulence in short time intervals.

3.10   Exercises

Problem 3.1 Mark as True or False

(a) The Jacobian method is applicable to both discrete and continuous variate transformations.

(b) If the range of a random variable X includes the origin, we cannot use the transformation $Y = 1/X$

(c) Marginal distributions can be obtained from joint distributions

(d) Marginal distributions determine joint distributions only when variates are independent

(e) Joint PDF of random variables can be obtained uniquely from joint CDF

Problem 3.2 If $f(x, y) = c(x + y)$ is the joint PDF of 2 discrete random variables ($x = 1, 2, 3; y = 1, 2, 3, 4$), find the constant c and hence obtain the conditional distribution of Y given $X = k$, and the distribution of X^2.

Problem 3.3 If $f(x, y) = c(x^2 + xy/2)$ is the joint PDF of continuous random variables where $x \in (0,1)$, $y \in (0,2)$ find (i) constant c, (ii) $P(Y > 1 | X > 1/2)$ (ii) $E(X)$ and $E(Y)$.

Problem 3.4 Find the unknown K in the following joint PDFs: (a) $f(x, y) = Kx^2y, 0 < x < 1, 0 < y < 2$ b) $f(x, y, z) = K(x + 2y + 3z)$ for $0 < x < 1, 0 < y < 2, 0 < z < 1$.

Problem 3.5 Let X and Y be the Cartesian coordinates of a point uniformly chosen in a unit circle of radius 1 centered at the origin. Find the distribution of $\sqrt{X^2 + Y^2}$.

Problem 3.6 If X and Y are IID exponential random variables with joint PDF $f(x, y) = \exp(-(x + y))$ for $0 \le x, y \le \infty$, find $P(X > Y)$. Find covariance (x, y).

Problem 3.7 What is the Jacobian of the transformation $u = aX + bY$, $v = cX - dY$. If X and Y are triangular, find the distribution of U and V.

Problem 3.8 If X and Y are IID $U(0, 1)$ distributed, find the distribution of $W = X/(X + Y)$.

Problem 3.9 If $f(x, y) = C/[x^k y^{k+1}]$ for $x, y \ge k$, find C and the marginal distributions.

Problem 3.10 Let X denote the number of COVID patients admitted to a hospital in a month which has a Poisson(λ) distribution. Among those admitted, let $Y_k = 1$ if patient k was fatal and 0 otherwise. If the total number of fatal cases per month is known to be binomially distributed with PMF $f(y|x) = \binom{x}{y} p^y (1 - p)^{x-y}$, $y = 0,1, \ldots ,x$, prove that the marginal distribution of total fatal cases is Poisson(λp).

Problem 3.11 If a pair of random variables (X, Y) has joint PDF $f(x, y) = Cx(b - y)$, $0 < x < 1$ and $0 < y < b$, find C. Check whether X and Y are independent. Find the conditional distribution of Y given $x > 1/2$.

Problem 3.12 Find the Jacobian of the *rotation* transformation $u = x \cos(\theta) - y \sin(\theta)$, and $v = x \sin(\theta) + y \cos(\theta)$.

Problem 3.13 What is the Jacobian of the transformation $x = r \cosh(\theta)\cosh(\phi)$, $y = r \sinh(\theta)$ $\cosh(\phi)$, $z = r \sinh(\theta)$?

Problem 3.14 Find the inverse mapping and Jacobian for the transformation $u = x + y + z$, $v = z - x - y$, $w = xyz$.

Problem 3.15 Find the inverse mapping and Jacobian for the transformation $u = x + y + z$, $v = xy + yz + zx$, $w = xyz$.

Problem 3.16 Find the Jacobian of the *shear* transformation $u = ax + y$, $v = y$. Use it to find the distribution of U when X and Y are IID (i) CUNI(0, 1), (ii) GAMMA(m_i,p), $i = 1$, 2, (iii) χ_n^2.

Problem 3.17 If X and Y are IID Rayleigh distributed, find the distribution of $Z = X^2 + Y^2$.

Problem 3.18 Find the Jacobian of the *rotation* transformation $u = x \cos(\theta) - y \sin(\theta)$, and $v = x \sin(\theta) + y \cos(\theta)$.

Problem 3.19 If X is Cauchy distributed, find the distribution of $Y = 2X/(1 - X^2)$.

Problem 3.20 If X_k's are n IID geometric RVs with parameters p_k, find the distribution of $Z = \min(X_k)$ (Hint: Chap 2., Example 2.9).

Problem 3.21 If X_k's are n IID geometric RVs with parameters p_k, find E[max(X_k)] (Hint: Use $E[\max(X_k)] = \sum_k E[X_k] - E[\min(X_k)]$) and $\min(X_k) \sim \text{GEO}(1 - \prod_{k=1}^{n}(1 - p_k))$.

Problem 3.22 If X_1, X_2 are IID N(0, 1), find the distribution of $Y_1 = \sqrt{-2\log_e(x_1)}$ $\cos(2\pi x_2)$ and $Y_2 = \sqrt{-2\log_e(x_1)} \sin(2\pi x_2)$ (the inverse being $X_1 = \exp\left[-\frac{1}{2}(y_1^2 + y_2^2)\right]$ and $x_2 = \frac{1}{2\pi}\arctan(y_2/y_1)$).

Problem 3.23 If X and Y are independent Cauchy distributed, prove that (i) $Z = (X + Y)/2$, (ii) $(X - Y)/(1 + XY)$ are also Cauchy distributed.

Problem 3.24 If X and Y are independent Gamma distributed, prove that Z = X/ (X + Y) is Type I beta distributed.

Problem 3.25 Suppose that X~BINO(n,θ) where $\theta \sim$ BETA – I(a, b). Find the unconditional distribution of X, conditional distribution of $\theta|X$ and prove that E($\theta|X$) = (a+X) / (a+b+n).

Problem 3.26 Suppose two fair dice are tossed. Find the density function of (X1, X2) where X1 and X2 are the scores that show up.

Problem 3.27 If f(x, y) = $Ke^{-(aX+bY)}$, find K and obtain the PDF of X/Y and X-Y.

Problem 3.28 If X~ χ_n (i.e.: X~ $\sqrt{\chi_n^2}$) and Y~BETA($\frac{n-1}{2}, \frac{n-1}{2}$) is independent of X, prove that (2Y-1)X ~ N(0,1).

Problem 3.29 If X~ χ_m^2, Y~ χ_n^2 and Z~ χ_p^2 are IID, find the distribution of U = X/Y, V = (X + Y)/Z and W = X + Y + Z.

Problem 3.30 If X~ χ_{2m+2n}^2, and Y~BETA – 1(p, q) be independent, find distribution of XY and X(1-Y) where $|J| = (x - y)(y - z)(z - x)$.

Problem 3.31 If x, y, Z are IID CUNI(–1, +1), find the PDF of (i) XY, (ii) XY/Z, (iii) (X + Y)/(X – Y), (iv) (X + Y – Z)/(Y + Z – X)

Problem 3.32 If $X_1, X_2, \ldots, X_n$ are IID GAMMA (m,p) prove that $Y_n = \min(X_1, X_2, \ldots, X_n)$ is distributed as GAMMA(mn,p).

Problem 3.33 Express the Cartesian coordinates in terms of cylindrical and spherical polar coordinates.

Problem 3.34 If X and Y are IID exponentially distributed, find the distribution of X + Y and X – Y.

Problem 3.35 Find the Jacobian of the transformation
$x_1 = r \sin(\theta_1) \sin(\theta_2) \cdots \sin(\theta_{n-2}) \sin(\theta_{n-1})$, $x_2 = r \cos(\theta_1) \sin(\theta_2) \sin(\theta_3) \cdots \sin(\theta_{n-2})$ $\sin(\theta_{n-1})$, $x_3 = r \cos(\theta_2) \sin(\theta_3) \cdots \sin(\theta_{n-2}) \sin(\theta_{n-1})$, $x_{n-1} = r \cos(\theta_{n-2}) \sin(\theta_{n-1})$, and $x_n = r \cos(\theta_{n-1})$. What is the inverse mapping?.

Problem 3.36 If X and Y have joint PDF f(x, y) = exp(–x–y), x, y$\geq$0, find the distribution of X/Y and X + Y assuming independence.

Problem 3.37 If X, Y, U, V are IID normal distributions, then prove that $UV \pm XY$ is Laplace distributed.

Problem 3.38 If $f(x, y) = kxy$, for $0 < x < y < 1$, find k and the distribution of $U = X/Y$. Check whether X and Y are independent.

Problem 3.39 Find the Jacobian of the transformation $y_1 = \sum_{i=1}^{n} X_i/\sqrt{n}$, $Y_2 = (X_1 - X_2)/\sqrt{2}$, $Y_i = (X_1 + X_2 + \cdots + X_{i-1} - (i-1)X_i)/\sqrt{i(i-1)}$ for $i = 3,4,..n$.

Problem 3.40 If Y has a chi-distribution with m DoF, and Z is BETA–I$((m-1)/2, (m-1)/2)$ is independent of Y, then $(2Z-1)Y$ is standard normal.

Problem 3.41 Verify whether the copula remains the same when the marginals on the RHS are replaced by increasing functions of them.

Problem 3.42 In what situations are the copulae $C(X_1, X_2, \ldots, X_p) = C(X_{\sigma(1)}, X_{\sigma(2)}, \ldots, X_{\sigma(p)})$ where $(\sigma(1), \sigma(2), \ldots, \sigma(p))$ denotes a permutation of the indices $(1,2,\ldots,p)$?

Problem 3.43 If C() is the copula of (x, y), what is the copula of $(1-x, y)$ when X has support (0, 1), and $(X, 1-Y)$ when Y has support (0, 1)?. How many distinct copulas exist when there are 3 variables defined on [0, 1]?

Problem 3.44 If C() is the copula of a random vector $[X_1, X_2, \ldots, X_p]$, all defined on the unit interval [0, 1], what is the copula of $[1-X_1, 1-X_2, \ldots, 1-X_p]$? What is the copula of $[g(X_1), g(X_2), \ldots, g(X_p)]$ where g() is a decreasing function?.

Problem 3.45 Let X and Y IID $N(0,\sigma^2)$, and $Z = XY/\sqrt{X^2+Y^2}$ and $W = (X^2 - Y^2)/\sqrt{X^2+Y^2}$ prove that the joint PDF of Z and W is $f(z,w) = 1/(2\pi\sigma^2)\exp(-(z^2 + w^2)/(2\sigma^2))$.

Problem 3.46 If U is any continuous distribution with support [0, 1] find the Copula of (U, $1-U$).

Problem 3.47 Prove that any bivariate copula satisfies the relation $\max(0, u+v-1) \leq C(u, v) \leq \min(u, v)$.

Problem 3.48 The personal income of a group of persons are known to be distributed exponentially with CDF $F(x, m, \theta) = 1 - \exp[-(x-m)/\theta]$ for $x \geq m \geq 0$, $\theta > 0$ where m is the minimal income in that group. If the marginal income distributions at various regions are known, explain how the joint income distribution can be obtained using copula theory.

Problem 3.49 If U = max[$X_1, X_2, \ldots, X_p$] and V = min[$X_1, X_2, \ldots, X_p$] prove that the bivariate copula of U and -V is given by C(U,-V) = max($u^{1/p} + v^{1/p} - 1, 0)^p$ which is the Clayton copula with parameter $\theta = -1/p$.

Problem 3.50 Check whether the function C(u, v) = $f(f^{-1}(u) + f^{-1}(v))$ is a copula where f() is any invertible function that maps [0, ∞) to [0, 1] with f(0) = 1, f(∞) = 0.

Problem 3.51 Prove that in two-dimensional case the survival copula is given by $\overline{C}(u, v) =$ u + v − 1 − C(1 − u, 1 − v).

Problem 3.52 If a trivariate copula is given by C(u_1, u_2, u_3) = $u_1 u_2 u_3 [1 + \theta(1 - u_1)$ $(1 - u_2)(1 - u_3)]$, check whether the variates are pairwise independent.

References

Allen, D. E.; Ashraf, M. A.; et al. (2013). *Financial dependence analysis: Applications of Vine copulae*, Tinbergen Institute Discussion Paper, No. 13-022/III, Amsterdam (http://www.tinbergen. nl), https://www.econstor.eu/bitstream/10419/87566/1/13-022.pdf

Annavajjala, R., Chockalingam, A., Mohammed, S.K. (2010). On a ratio of functions of exponential random variables and some applications, *IEEE Trans. on Commu.*, 58(11), 3091–97. https:// ieeexplore.ieee.org/abstract/document/5590326

Arendarczyk, M., et al. (2018a). Joint distribution of the sum and maximum of Pareto distributions appear in reliability theory, *J. Mult. Anal.*, 167, 136–156. https://www.sciencedirect.com/science/ article/pii/S0047259X16302482, https://doi.org/10.1016/j.jmva.2018.04.002

Arendarczyk, M., et al. (2018b) The joint distribution of the sum and the maximum of heterogeneous exponential random variables, *Stat. Prob. Lett.*, 139, 2–18, 10–19, https://doi.org/10.1016/j.spl. 2018.03.013

Au, C. & Tam, J. (1999). Transformation of variates using Dirac general function, *The Am. Stat.*, 53(3), 270–272. https://www.jstor.org/stable/2686109

Bailey, R.W. (1992). Distributional identities of beta and χ^2 variates: A geometric interpretation, *The Amer. stat.*, 46(2), 117–20. https://doi.org/10.1080/00031305.1992.10475864

Bansal, N., Hamedani,G.G. et.al. (1999). Some characterizations of the normal distribution, *Stat. Prob. Lett.*, 42, 393–400. https://www.sciencedirect.com/science/article/pii/S0167715298002351, https://doi.org/10.1016/S0167-7152(98)00235-1

Baringhaus, L., Henze, N., Morgenstern, D. (1988). Some elementary proofs of the normality of XY/($X^2 + Y^2)^{1/2}$ where X and Y are normal, *Comput. math. and its appl.*, 15, 943–44. https://www.sciencedirect.com/science/article/pii/0898122188900387 https://doi.org/10. 1016/0898-1221(88)90038-7

Beylkin, G. Monzon, L., Satkauskas, I.(2019) On computing distributions of products of nonnegative independent random variables, *Appl. and Compu. Harmonic Anal.*, 46(2), 400–416, https://www.sciencedirect.com/science/article/pii/S1063520318300228 https://doi.org/10.1016/j. acha.2018.01.002

Cadenza, M., Omey, E. & Vesilo, R. (2022). Revisiting the product of random variables, *J. Math. sciences*,267, 180–195, https://link.springer.com/article/10.1007/s10958-022-06123-0

Cacoullos T. (1965). A relation between t and F distributions, *J. of the Ameri. Stat. Asso.*, 60, 528-31, (also 60, p. 1249). https://doi.org/10.2307/2282687 https://www.jstor.org/stable/2282687

Cahillane, C. (2020). Sum of exponential and Laplace distributions, https://ccahilla.github.io/sum_exponential_and_laplace_distributions.pdf

Chattamvelli, R. (2024). *Correlation in engineering and the applied sciences: Applications in R*, Springer. https://link.springer.com/book/9783031510144

Chen, Y., Chen, D. & Gao G. (2019). Extensions of Breiman's theorem of product of dependent random variables with applications to ruin theory, *Comm. Math. Stat.*, 7, 1–23. https://link.springer.com/article/10.1007/s40304-018-0132-2

Cherubini, U. & Gobbi, F. (2013). A convolution-based autoregressive process, Chap. 1 in *Copulae in mathematical and quantitative finance*, Jaworski, et al. (Ed's), https://doi.org/10.1007/978-3-642-35407-6_1

Galambos,J. & Simonelli, I. (2004). *Products of random variables: Applications to problems of physics and to arithmetical functions*, CRC Press, https://www.routledge.com https://doi.org/10.1201/9781482276633

Gaunt, R.E. (2022). The basic distributional theory for the product of zero-mean correlated normal random variables, arXiv:2106.02897pdf

Groesser J. & Okhrin, O. (2021). Copulate: An overview and recent developments, *WIREs Comp. Stat.*, 1–22, https://doi.org/10.1002/wics.1557

Jones, M.C. (1999). Distributional relations arising from simple trigonometric formulas, *The Amer. statn*, 53(2), 99–102, supplement; 53, 393. https://doi.org/10.1080/00031305.1999.10474439

Kozubowski, T.J. & Nadarajah, S. (2010). Multitude of Laplace distributions, *Stat. Papers*, 51, 127–148. https://link.springer.com/article/10.1007/s00362-008-0127-2

Lun, Z. & Khattree, R. (2024). A general approach for imputation of non-normal continuous data based on copula transformation, *Commu. in Stat. - Simu. and Compu.*, 53(1), 567–594, https://doi.org/10.1080/03610918.2022.2025839

Ly S, Pho K-H, Ly S. & Wong W-K.(2019). Determining distribution for the product of random variables by using copulas. *Risks*, 7(1):23, https://www.mdpi.com/2227-9091/7/1/23. https://doi.org/10.3390/risks7010023

Ly, S., Pho, K-H., et al. (2019b) Determining distribution for the quotients of dependent and independent random variables by using copulas, *J. Risk and financial mgmt.*, 12(42), https://www.mdpi.com/1911-8074/12/1/42, doi.org/10.3390/jrfm12010042

McNeil, A. J. and Neslehova, J. (2008). Multivariate Archimedean copulas, d-monotone functions and L_1 norm symmetric distributions, Annals of Statistics, 37(5B), 3059–3097, arXiv:0908.3750pdf https://doi.org/10.1214/07-AOS556

Mekic, E., Mihajlo C. S., et al.(2012). Statistical analysis of ratio of random variables and its application in performance analysis of multihop wireless transmissions, *Math. Problems in Engg.*, 1–10. https://www.hindawi.com/journals/mpe/2012/841092/

Quine, M.P. (1994). A result of Shepp, *Appl. maths lett.*, 7(6), 33–34, Elsevier. https://www.sciencedirect.com/science/article/pii/0893965994900892 https://doi.org/10.1016/0893-9659(94)90089-2

Shepp, L. (1964). Normal functions of normal random variables, *SIAM rev.*, 6, 459. https://epubs.siam.org/doi/10.1137/1006100 https://doi.org/10.1137/1006100

Index

© The Editor(s) (if applicable) and The Author(s), under exclusive license to Springer
Nature Switzerland AG 2024
R. Chattamvelli and R. Shanmugam, *Random Variables for Scientists and Engineers*,
Synthesis Lectures on Engineering, Science, and Technology,
https://doi.org/10.1007/978-3-031-58931-7